PRACTICAL FORMWORK
and
MOULD CONSTRUCTION

SECOND EDITION

J. G. RICHARDSON, F.I.W.M.

*Lecturer, Cement and Concrete Association,
Fulmer, Bucks., England*

APPLIED SCIENCE PUBLISHERS LTD
LONDON

APPLIED SCIENCE PUBLISHERS LTD
RIPPLE ROAD, BARKING, ESSEX, ENGLAND

First edition 1962
Second edition 1976

ISBN: 0 85334 629 1

WITH 86 ILLUSTRATIONS
© APPLIED SCIENCE PUBLISHERS LTD 1976

All rights reserved. No part of this publication may be reproduced, stored in a retrieval system, or transmitted in any form or by any means, electronic, mechanical, photocopying, recording, or otherwise, without the prior written permission of the publishers, Applied Science Publishers Ltd, Ripple Road, Barking, Essex, England

Composed by Eta Services (Typesetters) Ltd., Beccles, Suffolk
Printed in Great Britain by Galliard (Printers) Ltd., Great Yarmouth, Norfolk

ACKNOWLEDGEMENTS

The writer wishes to thank the following for assistance with illustrations:

Balfour Beatty, Figs. 4.1, 6.1.
Cement and Concrete Association, Figs. 4.2, 6.3, 6.4.
Concrete Society, Figs. 5.1, 5.2.
Cubitt, Drake & Scull, Figs. 4.3, 11.2, 12.3.
Ductube Co. Ltd., Figs. 8.3, 8.4.
J. Laing Construction Company, Figs. 2.1, 10.10.
Rapid Metal Developments Co. Ltd., Figs. 10.8, 10.9.
Scaffolding (Great Britain) Ltd., Figs. 3.2, 4.4.

PREFACE TO SECOND EDITION

During the years that have elapsed between the publication of the first edition and this edition, the author has come to realise the need for specialisation particularly where such a broadly based topic as formwork is concerned. Each person is an expert in his own particular field, be it where a person is employed on a simple column and beam form system as constructed on site, or where another person works on a massive section of *in situ* concrete as required within a heavy civil engineering project.

Materials technologists, mechanical designers, formwork and falsework engineers as well as manufacturers and users of formwork, all seek at some time some particular information regarding formwork design, production and usage.

Thus whilst the author has set the scene and described formwork and mould arrangements within his own personal experience, he has also invited contributions from selected specialists on topics that include slipforms, special formwork, plastics and form construction.

In this way the scope of the book has not only been enlarged but has been updated. The original edition found acceptance among the more practically minded of those concerned with forming and moulding concrete, and it is hoped that this edition will continue to maintain that trend.

During the course of this revision the author has taken the opportunity to provide new and different illustrations related to formwork construction and practice and has added a bibliography, thus presenting the reader with an opportunity to increase further his knowledge of this important subject.

CONTENTS

PREFACE TO SECOND EDITION vii

CHAPTER 1

DEFINITION OF DESIGN AND THE APPROACH TO FORM AND MOULD CONSTRUCTION

1.1 Introduction 1
1.2 The architect 4
1.3 The engineer 6
1.4 The contractor or supplier and his staff 7
1.5 Sub-contractors and trades personnel 9

CHAPTER 2

PLANNING AND CONTROL OF FORM AND MOULD DESIGN

2.1 The formwork consultant 11
2.2 The contractor's design department 12
2.3 Design carried out on site 14
2.4 Mould design 15
2.5 Sub-contractors 15

CHAPTER 3

THE FACTORS THAT GOVERN FORM AND MOULD DESIGN

3.1 Specification and surface finish 17
3.2 Concrete considerations 19
3.3 Formwork construction 20
3.4 Placing considerations 22
3.5 Accuracy 24

3.6 Striking and demoulding 26
3.7 Striking times 28
3.8 Handling considerations 31
3.9 Mould requirements 32

CHAPTER 4

PRELIMINARY CONSIDERATIONS GOVERNING CONSTRUCTIONAL DESIGN

4.1 Interdependence of activities 35
4.2 Precast activity relationships 36
4.3 Planning arrangements 37
4.4 The activities of the formwork designer 37

CHAPTER 5

BASIS OF MODULAR DESIGN APPLIED TO MOULD AND FORMWORK

5.1 The optimum profile 43
5.2 Lifts and bay arrangements 46
5.3 Panel details 47
5.4 Vertical form section 49
5.5 Continuity and support 50
5.6 Precast concrete—mould design 52
5.7 The basic profile 54

CHAPTER 6

DETAILING THE FORMWORK SYSTEM

6.1 Choice of material 58
6.2 Provision of drawings and details 58
6.3 General assembly drawings 61
6.4 Sectional details 61
6.5 Panel details 64
6.6 Bay layouts and beam details 65
6.7 Sketches and schedules 67
6.8 Mould details 68

Chapter 7
FORM AND MOULD MATERIALS 1—PURPOSE MADE AND PROPRIETARY STEEL FORMWORK

7.1 Steel formwork 72
7.2 Sheathing arrangements 74
7.3 Adjustable steel props 75
7.4 Telescopic centres 77
7.5 Tie arrangements 78
7.6 Release agents 79
7.7 Retarders 81

Chapter 8
FORM AND MOULD MATERIALS 2—TIMBER, TIMBER-DERIVED MATERIALS, CONCRETE AND PLASTICS

8.1 Timber 83
8.2 Plywood 87
8.3 Hardboard 90
8.4 Particle board 91
8.5 Plaster and concrete 91
8.6 Plastics 92
8.7 Alloys and other materials 94
8.8 Formers for hollows and through-holes . . . 95

Chapter 9
THE MANUFACTURE OF FORMWORK AND MOULDWORK

9.1 Timber mould and form manufacture 98
9.2 Steel mould and form fabrication 104
9.3 Plastics working 107

Chapter 10
CONSTRUCTION OF *IN SITU* WORK 1—FOUNDATIONS AND WALL FORMATION

10.1 Foundations and ground beams 108
10.2 Kickers 112
10.3 Column bases 115

10.4	Ducts and pockets	115
10.5	Fixings and fastenings	117
10.6	Wall forms	119
10.7	Single-sided walling	128
10.8	Stairs and landings	130
10.9	Wall construction—general details	132
10.10	Circular and conical walling	135

Chapter 11

CONSTRUCTION OF *IN SITU* WORK 2—COLUMN, BEAM AND SLAB FORMATION

11.1	Column forms	140
11.2	Beam forms	149
11.3	Stair formwork	154
11.4	Standing supports	155
11.5	Reinforced concrete slab formwork	159
11.6	Stopends	163

Chapter 12

CONSTRUCTION OF MOULDS FOR PRECAST CONCRETE

12.1	General considerations	167
12.2	Precast piles	168
12.3	Floor units and decking panels	171
12.4	Moulds for mass production	173
12.5	Moulds for frame components	175
12.6	Mould construction—general	176
12.7	Stopend construction	181
12.8	Bridge beam moulds	183
12.9	Stack casting	184
12.10	Ducts for tendons	185
12.11	Precast stairs	186
12.12	Textured finishes	186
12.13	Ducts, culverts and subways	188
12.14	Precast products	189
12.15	Tilting frame manufacture	189

Chapter 13

CONSTRUCTION OF FORMWORK FOR SPECIAL APPLICATIONS, ARCHITECTURAL FEATURES AND SCULPTURE

13.1	Sculpture and low relief work	191
13.2	*In situ* moulded surfaces	192
13.3	General considerations	193

Chapter 14

MODERN STEEL FORMWORK SYSTEMS
(*I. Dunkley, Managing Director, Datron Gel Ltd.*)

14.1	Wallform	196
14.2	Tableform	198
14.3	Tunnel forms	202
14.4	Special forms	204
14.5	Precasting	206
14.6	General	207

Chapter 15

PLASTICS AS A MOULD MATERIAL
(*Peter J. Owen, Director, Bondeglass-Voss Ltd.*)

15.1	What does a plastics mould material offer?	208
15.2	Glass reinforced plastics (grp)	209
15.3	Storage of materials	209
15.4	Tools and equipment	210
15.5	Measuring equipment	210
15.6	Polyester resins	211
15.7	Catalyst (hardener)	211
15.8	Glassfibre	212
15.9	The master mould	212
15.10	Release agents	213
15.11	Mould reinforcement	213
15.12	Laminating	213
15.13	Usage	214
15.14	Release oils	215

15.15 Thermoplastics moulds 215
15.16 The method 216
15.17 Concrete 216
15.18 Two-part polyurethanes 217
15.19 As a mould material 217
15.20 Other plastics mould products 219
15.21 Expanded plastics 219
15.22 Materials 220

CHAPTER 16

SPECIAL STEEL FORMWORK—A CASE STUDY
(*P. R. Luckett, Technical Director, Stelmo Ltd.*)   225

CHAPTER 17

SLIPFORM
(*C. J. Wilshere, John Laing Design Associates Ltd.*)

17.1 Introduction 235
17.2 Detailed description 236
17.3 Structural design considerations 237
17.4 Equipment 237
17.5 Organisation 238
17.6 The start 239
17.7 Holes and pockets 240
17.8 Special aspects 240
17.9 Conclusion 240

CHAPTER 18

FORM CONSTRUCTION—TIE ARRANGEMENTS
(*C. J. Weller, The Sunderland Forge & Engineering Co. Ltd.*) 244

CHAPTER 19

THE ACTIVITIES OF THE FORMWORK DESIGN DEPARTMENT
(*K. Adams, Chairman, Joint Formwork Committee, Institution of Structural Engineers*)

19.1 Factors that dictate the decisions made . . . 256
19.2 The importance of attention to detail . . . 257

19.3	Preparation of drawings	258
19.4	Floor slab construction	259

Appendix
GROUP EXERCISES ON FORMWORK TOPICS

Formwork exercise I	263
Formwork exercise II	273
Planning a concreting operation	286
BIBLIOGRAPHY	287
INDEX	289

Chapter 1

DEFINITION OF DESIGN AND THE APPROACH TO FORM AND MOULD CONSTRUCTION

1.1 INTRODUCTION

In dealing with the design and construction of moulds and formwork for reinforced concrete it becomes necessary at an early stage to define the word design. It is not the intention of this book to instruct the reader in pure mechanical design, product of formulae and calculation, the fine approach of the engineer to problems of forces and moments, but rather to deal with the practical design of a mould or form for a given application. A design should be based on the constructional aspects, the choice of materials, the methods of incorporating them into moulds and the form systems in such a way that the materials fulfill the many requirements of the specification and the practical demands of use and re-use.

While the preparation of a design is frequently the province of the drawing office staff, much useful constructive design work is carried out on site or in the works itself. Practical arrangements for a mould or unit are frequently devised by those people who are engaged in utilising an existing system, with a view to improving methods employed in later stages of a project or on subsequent contracts.

At the design stage, when consideration is being given to the practical methods and the construction of the mould and formwork components, a bad decision may result in poor surface finishes and deformed concrete or units, or possibly damaged moulds, any of which can prove expensive if not dealt with at an early stage in the course of construction. Good mould and formwork design allows sound construction of a concrete structure or unit. Lack of attention to details such as striking or removing forms from concrete, or concrete from a mould, can destroy all the care and attention that has been expended on the gauging of the concrete mix, or the placing and compacting of the concrete within the form. A badly designed mould or form, although only a small part of the whole contract can, through lack of care, mar the completed work and upset the results of many other trades.

When sufficient attention has been given to the construction of a mould or system of formwork, and once the initial problems that can arise out of the first uses of what is probably a new and unusual approach to a problem have been solved, the mould or form system can be organised into a series of economic re-use operations. These operations can be designed to be carried out by operatives trained in a method of working in such a way, that the requirements of all who are concerned with the finished product are satisfied.

Provided that the construction is sound, and the members employed are mechanically correct, the system should lend itself to refinement and modification so that a complete form or mould system will have been attained.

In the succeeding chapters the many interesting facets of form construction will be examined, and a variety of methods of casting concrete will be discussed, although, of course, they present but a small proportion of the methods available to the constructor. It is the remarkable scope for invention and ingenuity, with rarely the same problem arising in the same way in similar work, that provides the interest in the subject.

A remarkable diversity of opinion is expressed regarding methods of carrying out the various operations and it is interesting to note that though there are certain accepted practices and specified methods, the approach to a given problem varies from person to person and from manufacturer to contractor. This book will describe accepted practice and mention possible innovation.

Craftsmen dealing with a trade use their knowledge of construction principles and skills which have been developed during years of fellow tradesmen's experience. As yet in mould and formwork there is little by way of a Code, and there are few set solutions to which reference can be made and carried into practice. Mould and formwork is a modern craft in which large volumes of plastic material are moulded and formed while hardening takes place and the material takes on a structural capacity, a comparatively new problem to those involved in building and civil engineering operations. The casting of metals has been carried out for centuries, the principles involved in providing a negative form or mould being shared with concreting. Here the similarity ends. The casting of steel or alloys has generally been regarded as a means of providing an approximation of a given shape for subsequent fettling.

In some instances expensive patterns are used and a fair copy of the given pattern is produced in the casting. In concrete work, however, it is normally impracticable to employ a pattern, the negative mould being constructed in the first instance and tolerances are laid down for the finished product. The work is frequently left with the face finish imparted by the mould or formwork thus providing a direct reflection of the ability and craftsmanship of those concerned in constructing the mould. The art of forming concrete is an exact one as exhibited in many structures, particularly those involving modern precast and prestressed concrete where individual units are cast to extremely fine limits.

The problems which provide the interest to the subject are rarely insurmountable and can, with the observance of sound construction principles, be solved in diverse ways. No writing on the subject should be considered as being complete since the ultimate in method to one person may prove to be the starting point from which another may begin to evolve a completely different solution or way of approach. Any one of a dozen methods while efficiently meeting the requirements of a particular authority, can also be more economical or quicker to construct.

This book can only outline the observations of the author which have been made while engaged on several sections of a lively and growing industry. By making clear the reasons for the adoption of any of the methods described, it is intended that a basis should be presented for sound mould and form construction and planning.

In order to consider the problems presented by a particular casting operation or the detailing of a system of mouldwork or formwork it is necessary to consider the authorities that govern the execution of the work. Each professional member or tradesman has a particular approach when laying down his requirements, whether it be by drawing or writing of a specification, and these must be borne in mind when the casting methods and formwork systems are designed.

Fig. 1.1. All the skills of the formwork designer and formwork carpenter are called into play when geometrical work and special features are incorporated by the architect and engineer.

1.2 THE ARCHITECT

The architect is primarily concerned with the aesthetic aspects of any project. Until quite recently, concrete was regarded as a material economically suited to the formation of foundations or heavy structural work, which could be suitably concealed by facings or dressing. Now the architect uses the material in functional and decorative roles, and frequently directs that the concrete be left as struck from the moulds or forms to provide some desired effect. The architect may express large areas of concrete in walls and columns, or alternatively he may incorporate a light precast and prestressed concrete skeleton, which can provide a delicate supporting medium for modern lightweight flooring and cladding systems. He can demand glasslike finishes to the concrete face or take advantage of the techniques of exposed aggregate work to provide ruggedly textured finishes. Concrete is essentially a functional material and it is not unknown for architects to require that the marks produced by the casting face of the mould or form be emphasised to add to the functional effect resulting from the use of large areas of the material. A smooth, even texture or featured surface may be specified to offset some other material incorporated in the design. Architects frequently turn to concrete to take advantage of the ease with which it can be moulded to provide the shapes and profiles of contemporary design, quite apart from its pure structural value.

In modern commercial architecture, concrete is the foremost constructional material and generally the specification will call for high quality surface finishes, which can be produced by the use of wrot timber, or ply sheathing or clean steel panels that constitute the face materials of the mould or form. Where large areas of concrete occur such as in silos, chimneys and similar structures, the architect may well call for modular features to be applied to, or recessed into the face of the concrete. Such features while undoubtedly complicating the formwork do much to enhance the appearance of the structure. Where decorative panels are required or exposed aggregate finishes are specified a variety of techniques can be used to express the exotic aggregates used within the concrete mix. Grit blasting, the use of retarders and washing techniques will still demand excellent formwork in the initial stages of construction. Precast concrete walling panels which are employed as permanent formwork may be used to enhance the colour and architectural form in the concrete structure.

In the design of schools and public or industrial buildings, architects are now taking advantage of the shell forms of concrete construction to provide large areas which are uninterrupted by supporting columns. Lattice frame forms of precast and prestressed concrete construction also achieve the same result. Concrete structures can be designed which give scope for large areas of glazing or light screening, and the degree of accuracy of the structural components obtainable through the use of carefully constructed forms and moulds are obviously of great interest to the architect. Modular cladding and glazing units can only be used successfully when the concrete structure conforms accurately to the drawings and specification.

The architectural concept calls for form and mouldwork that is constructed in a manner capable of expressing the architect's ideas of form and texture in the finished concrete work. A major factor underlying all these considerations is the need for the economic provision of a structure to meet the requirements of his client. In many instances an architect needs guidance on matters relating to concrete technology and the types of finish which can be achieved in given situations. The architect may wish to express some particular aspect of concrete and therefore co-operation will be necessary at all levels to ensure that the suitable effects are achieved.

1.3 THE ENGINEER

The engineer is concerned with the accuracy of the structure or the precast and prestressed components of a structure. Carefully designed profiles which are the results of design and calculation will have been laid down, and the engineer demands that these be formed within specified tolerances. The concrete mix must be carefully regulated and the specification must stipulate the engineer's requirements regarding the placing of concrete in order to avoid separation and to achieve consistency of compaction and density.

With these requirements in mind it is necessary to make provision for the vibration of the mix to ensure satisfactory compaction and so provide a durable component. Accuracy of profile is particularly important where the slender proportions of precast and prestressed concrete are concerned, while slight inaccuracies in mould profile can cause differential camber, deformed units and possibly render a unit structurally useless. The engineer requires that the formwork system be capable of providing concrete surfaces that are free from irregularities which can be brought about by deflection of the face of the mould or form. He will specify limits of deflection which are allowable within a given span, the quantity of concrete to be cast in one operation, and he will define the position of all construction joints. In post-tensioned work the engineer may well specify that the end blocks of units that contain the cable anchor arrangements, should achieve a given strength before the units are actually subjected to stressing.

Definition of Design and the Approach to Form and Mould Construction 7

While most specifications that govern concreting make the removal of the forms the responsibility of the contractor, the engineer will almost certainly insist that the formwork remain in position for specified periods depending on the nature of the work. He will need to be satisfied on such items as the curing of the concrete, protection from adverse weather and the avoidance of damage due to frost. Just as striking times vary on individual contracts so do the opinions of engineers on the most suitable striking times. These of course depend on the hardening qualities of the mix and the structure under consideration.

It is the engineer's responsibility to provide the working drawings for all concrete structures and components, while the architect's drawings are useful in showing general assembly details and indicating the surface finishes to be used in conjunction with the Bill of Quantities. The engineer's drawings should give details of profile, construction, joint positions and lines and levels thus providing the basis for all mould and form construction details. The engineer can be helpful in matters where small alterations to profile and design assist the contractor in achieving a better job especially with regard to visual aspects, or where problems of concrete placing occur. In all cases, the value placed upon good communication cannot be overstressed. The formwork designer must, when preparing his details, critically examine every drawing for the structure. It is during this process that discrepancies may come to light so that in the course of resolving these points a good relationship can develop between both parties. Small points such as these, dealt with in a constructive manner, will do a great deal to improve the performance on the contract, and will speed up the completion of the work by the various parties concerned.

1.4 THE CONTRACTOR OR SUPPLIER AND HIS STAFF

The contractor or manufacturer regards the provision of moulds and formwork as a means to providing a finished contract, and an item of plant or equipment to be used to present an economically successful method of retaining and casting concrete to programme. The contractor knows that on structural concrete work the appearance of the concrete produced from the system of formwork can have a considerable effect on his subsequent selection for invitation to tender for similar work. With the increasing emphasis in the civil engineering and construction industry on planning, prefabrication and programming, where there is any large percentage of concrete work in the structure, the form or mould system gives scope for the regulation of general site progress. The judicious provision of the correct quantity of mould and form equipment, combined with concreting plant of suitable mixing and placing capacity, is one of the major decisions to be made in the planning stage on the modern construction contract.

The contractor's agent or general foreman responsible for a given contract is interested in the practicability of the proposed system of formwork. Although he is responsible for maintaining the programme laid down for the concrete work, and also for satisfying the requirements of the architect and engineer with respect to the quality of concrete and the finishes obtained, he will be particularly concerned with the quality and quantity of material which is to be made available for use during construction. The agent or foreman's task is to set up a labour force capable of handling formwork on site in the most economical fashion. He is also responsible for the phasing of the erection, the concreting and striking work so that constant, gainful employment is provided for the other trades on site.

The agent or general foreman frequently has the assistance of site engineers who supervise the setting out and accuracy of the forms, while a foreman carpenter directly supervises the tradesmen and labour employed. The trades foreman and engineer are concerned with interpreting every detail from the structural drawing into the shape of the assembled formwork. The tradesmen acting under the trades foreman's instruction are employed on the tasks of fabricating the individual form panels. The trades foreman's counterpart in precast working—the mouldshop foreman—is likewise responsible for producing the moulds for each section of the work, or type of unit specified. Site carpenters' work is often be carried out under the worst possible weather conditions, often with little space or facility. In these situations they may well be called upon to produce complicated form arrangements, often to verbal instruction and rough sketches. When the moulds and forms have been fabricated the site carpenters erect and strike the system as dictated by the programme, working in conjunction with the other trades. Their main task is to maintain the forms between uses and alter and adjust for re-use as required. The carpenter's approach is mainly concerned with the maintenance of a repetetive sequence of operations, often with bonus targets in view. He requires constant instruction and information so that he can maintain the progress necessary to provide continuity of work for following trades, and to ensure that the accuracy is constantly monitored.

The mouldmaker works under shop or factory conditions usually direct from the engineer's unit drawings working to the instructions of the mouldshop foreman. It is his task to provide accurate moulds, soundly constructed, which are capable of withstanding multiple re-use imposed by modern precasting methods. He arranges for modification of moulds for subsequent re-use in producing the many types of unit cast from a given mould.

The viewpoints of the persons concerned with concrete form and mouldwork in civil engineering, building and precast concrete industries will vary but arising from the various considerations are points of approach in which all will share, namely: accuracy, considerations of cost and re-use factors, and the provision of facility for maintaining the required standard of concrete work with regard to placement, compaction and surface finish.

All these are capable of being controlled by the provision of well constructed mould or form systems.

It cannot be over-emphasised that the quality of the concrete with regard to surface finish and dimensional accuracy can have the utmost effect on the performance of a completed structure.

1.5 SUB-CONTRACTORS AND TRADES PERSONNEL

Formwork is a major item in concrete constructional work and once stripping of the forms has been carried out, many subsequent trades rely on the accuracy of the structural shell and suitability of the surface for the successful execution of their part of the contract. These tradesmen include:

    Masons for the cladding units
    Plasterers for coatings and renderings
    Heating, lighting and ventilating engineers
    Mechanical engineers and plant erectors

In fact the majority of trades rely on the ability of those who are concerned with the concrete structure at some time in the contract.

The quality of the structure or the precast or prestressed elements can be easily controlled at the time of manufacture. Imperfections can be eliminated and inaccuracies altered quite simply at the stage of form or mould inspection prior to casting. Failure at this time to carry out inspection work and modification of sizes can lead to great expense and can upset the programme at a later date.

Time allotted to inspection of mould or formwork following the erection or fabrication sequences comprise only a small item on the overall labour bill, but once concrete has been cast and matured this cost will cover only a fraction of the possible charges involved in remedial work necessary to provide the specified finish or allow the following trades to complete their work satisfactorily.

So far the requirements of a number of the people concerned in the production of concrete work have been discussed together with their approach to the subject of mould and formwork. Everybody will benefit from the production of concise drawings and specifications which have been prepared sufficiently early in the course of a contract. Planning can be carried out, programmes can be maintained and accurate good quality work can be ensured with reduced costs which result from an uninterrupted flow of work being made available to all concerned. It is the principles and practice of these activities which form the main subject of this book.

Chapter 2

PLANNING AND CONTROL OF FORM AND MOULD DESIGN

2.1 THE FORMWORK CONSULTANT

Contractors carry out the planning and design stages of the formwork for concrete structures in a variety of ways; these are largely governed by the scale of the contract and the organisation within their own contract departments. On large contracts such as those carried out for government departments, civic and industrial concerns a consulting engineer is usually nominated who is a specialist in form and mould design and who prepares all the schemes in connection with the concrete work. Such consultants employ a staff which is capable not only of calculating items of formwork and mouldwork but also of detailing the construction methods to be adopted. They design the formwork and detail such points as arrangements for handling the equipment during various stages of the contract. These consultants, by virtue of their constant engagement in large scale concreting practice, often throughout the world, provide highly efficient instructions to the contractor.

The consultant is well versed in the ways of carrying out the formwork, whether it be in steel or timber, or by composite arrangements of materials to satisfy the requirements of the client's engineer. He provides detailed drawings of the form systems for manufacture by the contractor and frequently arranges for the complete service of design and supply of the system, giving advice on its usage on site. The form engineer consults with the client's engineer over structural requirements, construction joints and the like and joins in negotiations between client's engineer and contractor, generally assisting in the control of the site work where mould and formwork is concerned.

2.2 THE CONTRACTOR'S DESIGN DEPARTMENT

The consultant form engineer is mentioned here as the model on which all departments, large or small, concerned with form design may well be based. The completeness of the service provided by his expertise is ideal, and a similar service provided by a section of a contractor's service department offers benefit to all sites and works with which the contractor is concerned. A large number of contractors now realise the value to be obtained from the form engineer's design department within their organisation. His department provide a facility which is capable of handling the mechanical design and preparation of construction details for the various mould and form requirements of the contractors' works. Under the auspices of the engineering department, or the contract manager, such a department does much to reduce overall mould and form costs. Immediately one person, or group of persons within the framework of a contracting concern becomes concerned with formwork procedures, uniformity of method results, thus providing opportunities for interchange and re-use of the form components. Many points are highlighted, items which would otherwise be missed due to the sectionalisation or isolation of the departments or sites concerned. Contractors find that savings can result from such minor items as the standardisation of thickness of sheathing and backing materials throughout the company, or the selection of one type of proprietary steel form system for use on all sites. Standardisation of such items, whether bought or hired, can result in mobility and flexibility of equipment and constant turnover of plant, with the knowledge that components transferred from site to site as work proceeds can be used satisfactorily with those already in existence.

In the same way standardisation of such items as tie rods or form bolt sizes can offer savings in both *in situ* and precast work. The contractor's design department can advise agents and foremen of the availability of mould materials for transfer as return loads on transport and routing to other contracts. Credits to first users of otherwise expendable materials may provide incentive for care of material, and the extra uses obtained by the company will further serve to reduce costs. The work of the contract supervisor can be reduced not only by a well organised formwork and mould department, but also by the assistance of site personnel in the preparation of details for mould and form fabrication. The preparation of cutting lists and the ordering of materials in bulk promotes savings, and allows for a monitoring process to be applied to purchases.

FIG. 2.1. The progress of the whole contract hinges on sound formwork. Here a circular ramping access road is being constructed in connection with a motorway. The correct positioning of the ducts for the prestressing tendons and the accuracy of the steel location depend on the accurate construction of formwork.

To ensure the success of a contractor's design department, it is essential that the design be carried out in a practical manner, with consideration for the conditions which exist on site when the system is in operation. The department, while dealing with various contracts, can do much to ensure that lessons learnt and successful methods of procedure developed on one contract can be made available to the staff on all other sites where concreting work is involved. All too frequently one contract can be proceeding profitably using tried and accepted methods or methods developed in the course of the contract modified into an economic routine while another contract being carried out by the same firm, but possibly under the control of a different supervisor, may be experiencing setbacks and difficulties. Such a state of affairs can easily be set to rights by the interchange of ideas, or even by the transfer of a number of key operatives from the site which is operating effectively, over a period of a few days during which they would be able to impart their knowledge of the system to those on the site previously disrupted by delay.

The staff of a contractor's design department must maintain a flexible outlook on job methods and be ready to accept suggestions emanating from contract staff and site personnel. While, generally, it will not prove economical to impose a radically revised method upon persons used to certain ways of working, it may well be that their own methods when tempered by fresh ideas can offer increasingly effective means of carrying out the work.

Many building and civil engineering contractors run their own temporary works and formwork design departments, and recently the trend is for such departments to offer their services to industry on a consultancy basis.

2.3 DESIGN CARRIED OUT ON SITE

A considerable amount of form design is carried out on site by junior engineers, section foremen and trade foremen. Much of this design is by nature extremely practical and considerable savings have been seen to result from the adoption of ideas and techniques generated at these levels.

Formwork provides an excellent area of responsibility for the young engineer and involves the processes of materials selection, structural calculation and decision-making which forms an essential part of engineering experience.

Experience gained in this way is doubly useful as the engineer has the opportunity not only of designing systems but of subsequently supervising the use of such systems and directly monitoring the results achieved.

It is essential that 'local' design arrangements should be supervised, not only to ensure structural integrity but to maintain an approach which provides a continuity of work and to standardise arrangements in such a way that the equipment ordered or produced is suitable for re-use or subsequent operations with the minimum amount of alteration and modification.

2.4 MOULD DESIGN

In the precast and prestressed concrete fields the planning of operations and the design of formwork is an important part of the production programme. It is therefore desirable that the responsibility for this work throughout the factory should be dealt with by one person or group of persons. In these days of highly technical procedures of mix control and prestressing, it is doubtful whether the shop or yard manager will have the time available to deal with the ordering or manufacture of mouldwork among his other duties. As moulds for precast and prestressed construction represent a considerable proportion of the cost of manufacture, design must be carried out to the benefit of the work in hand, and with a view to providing components and materials capable of re-use on succeeding contracts. As the robustness required to withstand constant handling, and the tolerances governing the work are often greater than those required on *in situ* concrete, the unit cost of mouldwork for precast and prestressed concrete exceeds the cost of the more complex arrangements of site formwork systems. The closest attention must be given to availability of moulds, provision of facilities for rapid modification (in order to produce varying types of unit), and the need to ensure daily castings or a regular cycle of operations occur within a works. It is the practice in many precast companies for the mouldmakers to be responsible for both mould design and construction. With the increasing technicalities of precast work, especially with regard to surface textures and tolerances, combined in some instances with stressing methods, it becomes more desirable that the persons responsible should have proper technical training combined with the knowledge gained from constant contact with manufacturing procedures.

2.5 SUB-CONTRACTORS

Contractors frequently employ specialist sub-contractors to carry out formwork erection and mouldmaking work and, while retaining control of the basic methods adopted, delegate the design and construction to these sub-contractors. The sub-contractor's service is usually built around experience gained from contracts carried out for many contractors on various types of work. The design of the systems used, and the methods of mould and form construction employed may be carried out by the sub-contractor's design staff working in conjunction with the main contractor's supervisor depending on the specific requirements of the contract, the aim being to ensure continuity of concreting for the main contractor.

Where the precast manufacturer is concerned, the need is for moulds to be suitably designed and constructed in order to offer the required number of uses at competitive prices. This is often achieved by sub-letting the work to a mould manufacturer who may work with steel, timber or composite materials. In this case the sub-contractor works from the engineer's unit drawings and the manufacturer's specification which governs the quality of finish required and the dimensional accuracy of the work. The precast manufacturer outlines the basic constructional methods to be adopted to ensure that the moulds provided are in accordance with the established standard methods of casting and handling used in his particular works. The mould thus manufactured must also be capable of being used in conjunction with the existing clamping equipment and grillages, and with the proposed methods of placing and consolidation of the concrete mix.

Whatever planning and control of formwork design and manufacture is adopted it is essential that the result embodies the latest developments of material and technique known to the contractor or manufacturer, and that the methods employed should be constantly reviewed as work proceeds in the light of the experience gained.

The present standards of concrete technology, the current knowledge of mould oils and parting agents, and the existing skills in mixing, placing and compacting concrete, demand that the formwork should be capable of imparting finishes consistently during the period of their use, in such a way that the completed component reflects these very high standards of skill.

Chapter 3

THE FACTORS THAT GOVERN FORM AND MOULD DESIGN

3.1 SPECIFICATION AND SURFACE FINISH

Amongst the finishes called for by current specifications is that in which concrete cast from the forms should present what is termed a 'high quality surface finish'. This type of specification is encountered where a structure is such that on completion all concrete is exposed to view. Such specifications demand careful choice of sheathing material for the mould or form face. This sheathing face is often treated with a sealer by brush or spray (as discussed later) to reduce the possibility of adhesion between the concrete fines and the form face. Plastics faced plywood is an excellent material for this type of work. Accurate jointing of the face material is however essential to reduce the risk of localised honeycombing through grout seepage, and foamed rubber and plastics strips will assist in the prevention of grout or paste loss.

While many contractors are at present able to produce excellent work which is accepted to be within the specification of 'high quality surface finish', the greatest care is necessary when erecting and striking forms for this type of work. Vibration of the concrete must be carefully carried out to avoid damaged or burned faces of forms, and the finished concrete face must be protected from damage during subsequent operations. A number of materials can be used to produce concrete of this standard though care must be taken that defects in the sheathing face do not impart unsightly marks on the face of the concrete. The slightest irregularity of the sheathing surface will tend to mould the concrete in such a way that the incidence of light on the face is altered locally resulting in an inconsistency of appearance.

Where sheet material is employed to line the form units, minute variations in the thickness of boards from a batch of sanded stock can present breaks in the face which are likely to mar what would otherwise be excellent work, and current practice tends to make use of very large sheets or continuous strips of plywood.

A more general specification is that in which 'fair faced' concrete is specified. A face which is the product of a prepared board, sheet material, steel plate or plastic sheathing described as 'wrot', 'prepared' formwork, or 'fair faced' is a poorly framed description and one that is not clearly understood. It is taken generally to indicate smooth and reasonably blemish-free concrete. For the majority of 'fair faced' concrete the contractor can satisfy the specification by using ply or steel-faced form panels or plywood for the mould or formwork sheathing, this being supported in such a way as to remain within the deflection tolerances. Douglas fir, Finnish birch or hardwood-faced plywood are usually the most acceptable materials for the purpose. Where the pattern of sheathing joints is maintained symmetrically, within a given bay or area of concrete, the engineer usually accepts modular steel or ply-faced panels for the sheathing within this specification. If there is an objection to the small panel outlines, resulting from the use of modular panels, it is possible to obscure them either by covering the joints in the form face with adhesive tape or by filling with plastic material or plaster to flush off the joints. A considerable improvement in the face can be achieved by the use of foamed plastic in the joints within the form face.

Frequently, an architectural feature or fillet at intervals on a 'fair faced' area of concrete is required, and these can be a means of cloaking joints in the form face. The formation of such features is discussed in a later chapter. For some classes of fair faced work, rubber sheeting, rolled or extruded to the required profile, can be used as the sheathing material. Chapter 8 deals with the use of rubber and plastics in formwork. Even cardboard, fluted hardboard and corrugated asbestos sheeting can be employed as facing to the form units to achieve special effects, coffers, recesses and similar details.

However carefully concrete is placed and consolidated, it is difficult to produce a lift of wall face without a number of pin holes or tiny voids occurring where entrapped air or water bubbles have been in contact with the form face. Often such pockets of air are brought to the face by the turbulent action imparted by the vibration of the concrete mix; absolute care is essential to avoid a slight loss of fines adjacent to the previous lift or casting of concrete, where form panels overlap to provide a rigid fixing. In both cases any dressing or reworking of the face must be by permission of the architect and engineer.

'Sawn formwork' can consist of any of the previously mentioned materials once the maximum permissible number of uses on fair faced work has been obtained. Sawn timber boards, asbestos sheathing, underlay sheathing grades of plywood, and in foundations corrugated steel sheet or polythene sheathing, may be used. Many contractors utilise single thickness brickwork as 'sawn formwork' to pile caps, column bases and ground beams, while concrete blockwork may provide a locally acceptable alternative. Above ground level steel plates or rough ply can be used though subsequent finishing operations, such as plastering, may require the removal of the nibs formed at the joints.

Surface textures are becoming more popular with architects who are specifying exposed aggregates and highly featured faces to gain effect. Exposed aggregate finishes can be obtained in a variety of ways. Grit blasting and the use of retarders on the form face are the two techniques most usually employed for both precast and *in situ* concrete. High labour costs now tend to make such indirect methods as the tooling, bush hammering and roller comb processes too expensive.

Where highly featured panels or details are required, plastic sheet and concrete surfaces can be used as the sheathing material, and as discussed in later chapters, careful selection and fabrication of materials should provide the required number of re-uses of the forms.

3.2 CONCRETE CONSIDERATIONS

The foregoing should have illustrated the manner in which a specification governs the choice of a sheathing material. In all cases the aim of the designer will be the formation of accurate profile with minimum deformation due to deflection or wear of the sheathing surfaces through constant re-use.

A great number of moulds or forms fail due to grout loss, thus construction must be detailed to ensure that a grout tight form or mould is produced. The form manufacturer's greatest concern is the retention of the fresh concrete fines which, particularly where a high degree of vibration is employed, infiltrate into and harden within the slightest open joint or intersection between form faces.

Cleaning of moulds may not effectively clear the fins of hardened concrete from open joints and junctions, and subsequent build up can cause distortion of the mould or form face, and inaccurate profiles and honeycombing. These defects often occur at critical points in the structure such as arrises and features which require dressing after removal of the formwork. Fabricated panels can easily be rendered useless by this infiltration action during succeeding operations and may even require rebuilding to provide further satisfactory castings. Accumulations of hardened fines on the face of forms or at joints in the construction can cause tearing of the concrete face during striking due to the key offered to the newly cast concrete by the irregular face of the concrete deposit. Open joints and intersections can be avoided by good construction, and in the case of steel formwork close joints can be obtained by regular wire brushing and treatment with mould oils and parting agents. The greatest care is therefore needed to avoid damage to the arrises during the striking operation or when panels are handled between uses. When panels become damaged or dented the marks should be beaten out or filled with a weld followed by grinding. When timber and plywood are being used, bad edges will need to be sawn back and built up by insertion of rips or 'joiners'. With timber and ply fabrication, the use of a good quality weather resistant adhesive in permanent joints helps to prevent grout infiltration. Casein glues present a fair durability when exposed to the weather and offer sufficient resistance to water to permit their use in mould and form construction, though the more expensive forms of adhesive such as urea/formaldehyde resin adhesives present joints capable of resisting long-term moisture attack. In the past glued and thicknessed boards were used prior to the availability of reliable plyboards and frequently gave up to 100 uses without breakdown of the joint, although the selection of casting method was critical in these instances.

3.3 FORMWORK CONSTRUCTION

The quality of construction is governed by the requirements of the specification for the concrete. It is rarely possible to obtain a good quality finish from cheap, short lived sheathing material or even by using excellent facing materials if the grillage or backing is not soundly constructed and accurately fabricated. It may well be that in the initial casting operations the finished face will be satisfactory, but with subsequent uses of the backing, joints may open up due to movement, or the face may become deformed by deflection of the framing. It is more economical to provide securely-built systems of backing or bearers which are faced with relatively dispensable materials suited to the number of uses required and the specification. In this way the forms may be refurbished when wear occurs, thus maintaining high standards of surface textures and finishes.

FIG. 3.1. Timber formwork being constructed in a works. The sub-structure is designed to take support from a tubular scaffold. In this way the geometric form is generated from a simple scaffold arrangement.

The limits of deflection and dimensional tolerances allowed by the specification dictate the construction employed. In precast and prestressed concrete work, supports can be placed as frequently as required and frequently grillages exist for this purpose. It may prove impossible in *in situ* work however to place props or supports in the exact positions required to provide the most economical layout of form framing. In the case of *in situ* work the problems associated with the formwork can become critical structural problems, and special construction and falsework will be required to span across the available supports.

Some proprietary systems do provide a special facility for large uninterrupted bays of formwork by the use of lattice-type units. These can be extended at will, and the relative lightness of the units makes them easy to handle. Tubular scaffold can be used to bridge over access ways for site traffic, or to span between sound footings such as those offered by column bases in bad ground with intervening dig or loose fill. Designers concerned with these problems should study the Institution of Structural Engineers/Concrete Society Technical Report on *Falsework* which covers these sort of problems.

3.4 PLACING CONSIDERATIONS

When pumping or pneumatically placing concrete by pipeline massive pressures may develop, where in order to avoid modifications to the length of pipe runs, concrete has to be constantly placed in one part of the form and flowed into position by vibration. Ready-mixed concrete placed by chute direct from the transporter can also present the same problems. The form or mould system will be designed for a given head of pressure and this can easily be exceeded, where for lack of insertion of an extra length of pipe or by hasty placement direct from the vehicle, the rise of concrete can occur too rapidly at one point in the form and thus exceed the designed rate of placing.

The form construction employed must be capable of withstanding the impacted loading brought about by the discharge of skips of concrete from a height either through poor liaison between concreting gang and crane driver or because of difficulty of skip access. Form arrangements must be designed to carry the loading which develops from railed conveyors or from stationary plant, or from belt action conveyors and the vibratory action resulting from the truck movement or motor drive. Preplanning of positions likely to be occupied by plant on the surface of floor formwork should allow for the provision of extra local supports. Where sites are congested, steel reinforcement may have to be stored immediately prior to fixing, and if hollow blocks are to be used for floor construction they may be bulk loaded on to the formwork before the steelwork is laid.

Formwork provides in many instances not only a sheathing for supporting concrete but also a working area for many other trades. Steelfixers require working space to assemble the cages of reinforcement; heating engineers will often be employed in installing heating grids, while mechanical engineers may well require the formwork to support heavy steel fittings which are to be cast later integral with the concrete. All these factors need to be considered while the supporting system is being designed since failure to do so could result in an improper assessment of the strength requirements of the form.

It thus follows that all mould and formwork should be so designed to facilitate the placement and vibration of concrete.

Apart from the normal openings at the head of a lift of concrete, or top of a concrete unit, it may be necessary to provide access for either poker vibrator shoes or fixing pads for rotary vibrators. While such openings are now covered in the specification for narrow walls, I-section beams and hollow units call for the provision of such openings. Walling and units which have cill shutter boards to assist in the formation of openings for windows or services demand special treatment to ensure a good fill. Board-width pockets used in conjunction with sheet-steel chutes help in the concreting of I-section beams, by allowing entrained air to escape by permitting the insertion of vibrators. When concrete begins to flow from such apertures it can be screeded level with the general sheathing, and then the cover plates inserted. Where open flanges are left in I-section beams a board provided at the root of the flange helps to prevent surge when the web section is first cast. This can also apply to trough members cast with the recess uppermost.

Engineers sometimes allow large I-section units to be cast on the lines of *in situ* practice, i.e. the bottom flange and kicker being cast first followed by the web and top flange. Where walls are cast in storey height lifts, the form panels can be arranged in such fashion that certain sections may be added as work proceeds, and the height of the concrete in the forms increases. This is particularly helpful where heavy reinforcement is employed to avoid conic bridging of the concrete over the reinforcing steel which can result in honeycomb or pocket formation. One face of a very tall column can be broken into suitable sections to allow insertion as concreting proceeds, though as with the other items mentioned, a tight joint is required to avoid local honeycombing and grout loss at this point. Where top

Fig. 3.2. Undercasting and the formation of hoppers presents problems of uplift on formwork. Note the twin members used to restrain the hopper forms, shown here in the foreground.

forms have to be provided for sloping surfaces such as those used for hopper formation or at the root of barrel roofs, an arrangement of panels or boards, similar to coal boards, placed as concreting proceeds helps in the placing and vibrating operations. The need for a top form on sloping surfaces is often a function of the mix design and workability.

3.5 ACCURACY

Accuracy of the mould or form is an essential factor which governs construction, and in this respect it can be argued that the dimensional accuracy called for in the majority of *in situ* concrete work is less exacting than that of precast concrete. Dimensions are applied to the face and profile of the concrete or unit regardless of cubic content. In practice it may be more difficult to obtain a high degree of accuracy where *in situ* masses are concerned due to the correspondingly larger units of formwork in use, since awkwardness of form handling limits the fine adjustment of the profile.

Site conditions often render accurate working more difficult to achieve than is the case with work carried out under factory conditions. Additional effort on the part of the designer can however assist in this matter which can be resolved by location and jigging.

The size of the formwork unit has less effect on the degree of accuracy of work obtained than does the section and geometric form. There can be little in mould or formwork harder to construct accurately than forms which are used for circular or flewing work. It is in such areas that the constructive skills of the designer can provide economies.

Wherever applicable mould construction and quality of materials utilised must be gauged to the requirements of the specification, though a further consideration will be that of economics with regard to plant and formwork equipment costs. It is always advisable to consider quality of construction or choice of system in the light of future work on succeeding contracts. By investing in standardised equipment the contractor will be able to ensure further returns from subsequent concreting contracts.

In recent years considerable work has been carried out with regard to constructional accuracy and permissible deviation for various classes of construction. The recently published report on *Formwork* by the Joint Committee of the Institution of Structural Engineers and the Concrete Society highlights the factors concerning quality standards and makes the point that there is a need to identify the appropriate standard of concrete finish and accuracy. The report comments on the increase in costs which will result from the specification of close tolerances. The report proposes that the normal standard of commercial concrete structures resulting from construction carried out to dimensions on drawings and in the specification should provide a 'norm', any excess or oversize concrete which does not impede subsequent fixing and fastening representing merely a materials cost to be borne by the contractor.

A further grade of workmanship is suggested: result of working to a set of values and applying inspection and checking at all stages.

The next grade of workmanship would be that which involves the application of special form materials, labour and supervision and comprehensive inspection and checking.

It may be that special tolerances are required for some particular purpose or use of a structure, and in this instance more elaborate methods would be used coupled with specialists involved in checking and inspection. Very special tolerances are rarely required and their adoption demands special expertise on the part of all concerned in the construction process.

The whole question of tolerance, accuracy and suitability of formwork has tended to be a rather 'grey' area since engineers and architects have been reluctant to specify, or in some places comment, on formwork arrangements lest by so doing they assume responsibility. The trends are now however towards the assessment and specification of realistic tolerances with a tendency to methods which specify the critical dimensions stating permissible deviations, yet allow items secondary to the function of the concrete to be constructed to reasonable or appropriate standards.

It should be remembered that specified tolerances generally relate to the completed concrete component, and allowances should be made when designing, fabricating and erecting formwork for movements, deflections and even the dimensional stability of the form components each of which may affect the dimensions of the finished concrete components.

3.6 STRIKING AND DEMOULDING

The units comprising a formwork system or mould must be capable of being removed from the concrete without damage to the structure or to the form unit. Where features are being formed which tend to trap the mould or prevent the self weight of the form from assisting in the striking process, the features should be rendered wholly or partly removable. Where units tend to bind between returns then striking pieces must be inserted either for removal prior to the striking operation or after the remainder of the formwork has been stripped from the concrete.

Engineers concerned with concrete structures and products appreciate the difficulties of forming square apertures and indents within the concrete face and generally allow the provision of a lead or draw in the profile to achieve clean striking of the mould unit without damage to the form face or arrises of the concrete. More form damage can be incurred through bad striking practice and incorrect storage between uses than that experienced during the actual forming of the concrete. Where it becomes necessary to use wedges or bars to remove forms from the concrete face, the design or construction of the mould system is suspect or possibly the actual striking action is being carried out at the wrong time. An initial adhesion has to be overcome which is caused by the need for air to percolate between the form face and concrete. Immediately this temporary bond has been broken the panel should be quite free to slide from its casting position ready for cleaning, oiling and re-erection. A form or mould unit should require a minimum amount of dismantling to facilitate its removal from the concrete.

With complex construction such as is used on precast or prestressed unit moulds it is essential to make full provision for the striking of deep feature formers. This can be achieved by means of built in arrangements of screw pads bearing on the concrete surface or pneumatic and hydraulic rams arranged to force the form faces evenly apart. In this way minute features imparted from the mould to the concrete are not damaged. Where soffit sheathing to concrete floors is concerned, 'crash' striking of formwork is often employed for reasons of economy. This process is not to be recommended and care is required to avoid form components being damaged through corners of panels striking the faces of other panels with resultant dents and scrapes which can spoil future work. Particularly in the case of special surfaces, the formwork system should incorporate means of lowering the joists and bearers to allow release of the sheathing which can then be carefully removed for cleaning prior to re-use.

When striking formwork, particularly that employed on multi-storey work or in the vicinity of site boundaries, the area should be lined out by spare joists previously stripped from inside bays to prevent materials from falling and thus causing accidents. Safety rails should be erected around openings prior to striking of the sheathing from below large apertures left in the slab, in order to provide safety for operatives employed on the upper floor.

A further consideration regarding the striking of formwork is the requirement that standing supports should remain in position once the main form faces have been removed. Re-shoring by replacement of props after completely striking the formwork from the underside of beams or floor slabs can be extremely dangerous. There is a possibility that the jacking action of the inserted props may introduce stresses which have not been catered for in the design of the structure. The commercial systems incorporate quick stripping actions using what are virtually double-headed props which are particularly valuable in this situation. The structural concrete remains continually supported by the main prop member while the secondary head is lowered for the removal of the sheathing and support system.

The whole question of prop easing and removal is now receiving considerable attention, particularly as it is thought that excessive stresses may develop in the standing support members in multi-storey work where the structure is unable to contribute to load bearing while in its shored state. It would appear that shores should, where succeeding floors are cast in rapid succession, be 'eased' to allow floor and beam members to deflect and contribute to the load bearing arrangement.

3.7 STRIKING TIMES

There is considerable difference of opinion over the time which should elapse between the casting of concrete and removal of the formwork and moulds. Recently, research and measurement has been carried out to establish the maturity of the concrete mass necessary to resist frost attack, and certain tables have been published which offer guidance on the matter.

With regard to columns, walls and beam sides these can generally be removed in normal conditions within 24 hours of the concrete being placed. It may be necessary where special finishes are specified to remove such formwork at a very early age and, provided that curing is continued, there seems little reason why forms should not be removed from five to seven hours after casting.

As an example stopends can be removed at two to three hours from casting in order that the surfaces can be simply prepared to enable a sound construction joint to be achieved. Again, curing must be continued after the concrete surfaces have been prepared.

In the case of beam and slab soffits it is usual to relate the striking time to the compressive strength of concrete cube specimens prepared for the purpose and cured under the same conditions as the structural component.

In general construction it is usual to provide standing supports, i.e. props which bear against the concrete which support a continuous feature that enable the main areas of sheathing to be stripped and re-used about three days after the slab has been cast or seven days after the beams are cast. The standing supports remain until the structural concrete has achieved an agreed percentage of the specified 28-day strength. Where rapid construction techniques impose the dead load of succeeding floors on a system of standing supports, these should be eased to allow the structural members to contribute towards sustaining the load and thus avoiding the situation where the props to the lower floors are over-stressed.

Striking must be carried out systematically and without undue force. The operation must be such that there is no damage to the concrete or the form and by such means as to avoid impact or sudden shock to previously cast parts of the structure. Thin wedges, jacking screws, compressed air and similar means of applying force must be used in preference to bars, hammers and similar tools. Even the swaying of a crane bond in an effort to break the form/concrete bond is sufficient to crack the concrete at the base of a wall. Crash striking where all the supports are removed and the sheathing allowed to drop is not only dangerous but can cause structural defects in both the newly-cast floor due to the temporary imposition of the dead weight of the system and the floor below, due to the impact of falling equipment. With regard to precast concrete components, units may be either instantly demoulded or within seconds of casting. Alternatively, units may be cast on a daily cycle and in this case care should be taken to ease moulds while there is still some degree of paste on the face of the units. This applies particularly to complicated units and units being cast from tip-out moulds or one-piece moulds. There have been recent instances where units of up to 40 tonnes mass have been cast from one-piece moulds with only a minimum amount of lead or draw. For high quality surfaces, whether precast or *in situ*, there must be complete uniformity of the striking and curing time. The longer the form remains in place the darker will be the concrete face and even hours of difference can affect the finish. There is evidence where the difference in striking times between succeeding lifts of concrete is visible some years after the contract has been completed and differences in rates of curing can be seen between adjacent boards of different absorbency many years after the work has been completed.

A formwork system must be capable of being struck from the concrete face and handled around obstructions such as scaffold frames. Form operations must be phased with scaffold erection to ensure freedom of movement for both trades. Where forms are used to produce chambers, or ducting with manhole access after concreting, the panel size must be regulated by the opening size that remains following the completion of all concreting and striking.

When formwork is being handled between uses by crane, strongbacks should be provided to avoid springing of the face. Bolts and ancillary parts of a form can be rendered captive on chain or wire and thus will travel with the form and save time in assembly. Washers, bolts and wedges collected in boxes at the time of striking should be ready for re-use at the time of erection. Forms should be fabricated in such a way that the minimum amount of bolting is required before the arrangement becomes self-supporting since this ensures the minimum demand on crane time as well as rendering the form safe from accidental displacement. While regulations do not always allow a working platform to be suspended from a form arrangement, hand holds for the use of operatives during erection facilitate the work. Brackets to support heavy walings while bolts are inserted help to make conditions safer and speed up the work.

It is essential when considering the striking method to be adopted that the following be considered:

Is the panel to be placed directly into its next working position? If not, is storage space available?

Can the cleaning and application of the parting agent be made during the operation of moving the form?

Are there local obstructions to stripping such as projecting bars or corbels?

Has access been allowed for the release of all bolts retaining cast-in fittings?

Has clearance been allowed to suit the tie method selected, i.e. between concrete face and scaffold arrangements?

In certain instances the conditions of access will have changed radically since the formwork was first erected. Furthermore, concreting or erection operations may have resulted in certain obstructions being introduced which precluded the use of the equipment used in the initial erection. An instance of this problem occurs where heavy centres have been simply lowered into place by crane to form a bridge soffit or some similar part of a concrete structure. At the time of striking there is now a major obstruction, to the removal of the centres, in the form of the newly cast deck or slab. Although this is a problem of crane usage, the problem can be magnified where manual removal of forms is rendered more difficult because of access problems, or even awkward working positions.

It may well prove worthwhile here to extend the runners or rails which support the centres so that they may be slid out from under the deck into a position where they can be readily lifted in batches by crane.

3.8 HANDLING CONSIDERATIONS

When units of formwork are to be handled manually throughout the course of a contract, it is advisable to restrict the weight of the individual units to that which may be handled by two men working as a team. If required, a third man can assist in the insertion of bolts which retain the panels in position. Three- or four-man lifts make for uneconomic handling due to the difficulty of obtaining satisfactory purchase for lifting. On high buildings or exposed sites, it is necessary to ensure that forms are restrained mechanically during lifting, or at the least tethered by wire strops to the scaffold or some other convenient anchorage. Gusts of wind can easily sweep men and units from the scaffold while their minds or balance are concentrated on the job in hand. The smallest block or bolt left loose during striking and erection can be a potential source of accident to those working on lower levels or pedestrians and motorists outside the site. The use of buried anchors and commercially available clamps helps to provide safe working since anchors ensure that immediately after the lifting action has been completed, the form can be fastened back to the concrete of the preceding lift. Form components can also be bolted individually into place to allow steel fixing and other erection activities to proceed.

The system provided to cast precast or *in situ* concrete must be mechanically sound and designed with regard to a number of factors which, while not lending themselves to classification in order of priority, will effect the design to varying extents over a range of types of work.

The strength requirements must not bring about unmanageable panels. A panel can be constructed to retain a given quantity of concrete efficiently, with the minimum of framing or backing material in its construction. Such panels however can prove quite unsatisfactory when struck from the concrete face and mechanical means are used to handle them. The advent of the tower crane as a standard piece of equipment on even the smallest of construction sites increased the amount of mechanical handling on building sites in much the same way as derricks, gantries and mobile cranes originally affected civil engineering work and precast work. Large mould and formwork panels can be produced and moved in mass between points of work but it is useless to design form units which are only capable of being moved using this equipment if at some stage in the work they require to be moved manually. The largest assemblies are best constructed by the use of several sub-assemblies linked by continuous framing members which may be removed at times of programme difficulty or delay. In these situations travellers also serve to uncouple formwork handling requirements from the remainder of the work.

3.9 MOULD REQUIREMENTS

The preceding points which relate to *in situ* work, equally apply to the factory or yard concerned with the casting of precast concrete. At the time of striking accessories should be placed into suitable containers or re-entered into the moulds until required for further use. Where ties can be re-inserted and the mould re-assembled after cleaning, oiling and modification these assist in ensuring that the mould parts are kept free of damage until next required for use. Formers and stooling pieces which produce vital recesses in the finished concrete product are unlikely to be removed or used on other work in hand thus avoiding subsequent loss or damage. The corners of features and stopend panels can easily be damaged by contact with other components, and such damage should be avoided when the whole unit is assembled for storage blocks. Where close casting of units on a precast bed is being performed and where gang moulds are in use, care must be taken to ensure that dowel bars and through ties can be withdrawn without fouling adjacent mould set-ups or completed units which have been left to cure. The mould sides for deep section I-beams require considerable space to allow the feature to be cleared from the concrete prior to being lifted for transport to the next pallet or casting position.

Where handling of forms and moulds in a works or yard is carried out by crane, and particularly where striking is to be performed by this means, suitable lifting eyes and spreader bars or strongbacks must be provided. Internal form units to gang-cast piles, for instance, require considerable force to be applied to release them from the concrete and substantial lifting eyes are needed when these forms are in constant use. On casting beds, moulds often suffer damage through careless stacking in gangways, or where they have been placed adjacent to access roads where cranes and dumpers work. The drivers of these cranes while concentrating on the lifting of a unit or on the loading of a lorry, may be unaware of damage that can or is caused. Piles of moulds or panels in access ways can cause the loss of many production hours in the course of a week, where operatives have to clamber around to carry out their work and such stacks may also cause a serious accident hazard. Moulds should be removed from the immediate working area of the bed between uses while unit removal and steel fixing is being carried out. Free moulds should be stacked on level bearers to avoid warp and wind, and care must be taken to avoid the denting of steel faces by impact. Timber forms should be hosed down in hot, dry weather to avoid shakes opening in the timber sheathing or joints opening between the boarding. A particular instance where attention is essential to the provision of mould storage between uses, is in the case of battery casting with removable divider plates. Divider plates or liner forms have a mass of many tonnes and tend to be placed or leaned against almost any convenient standard, whether it is an adjoining mould or a column in the casting shop. Moulds standing in other than purpose-made racks of a substantial nature form a considerable accident hazard as they require only vibration or the impact of a passing mass to dislodge them on to workmen who may be working on nearby moulds.

Small mould side members, particularly those of steel with heavy features, tend to be most unstable and while resting on the workshop floor can so easily turn, trapping an operative's foot or leg in the process.

The following points should be noted in handling moulds and components:

That no loose materials should be allowed to fall on to operatives as the moulds are moved (this also applies to dust which can cause eye accidents).

The need for workers to approach chains, slings and particularly spreader bars during the actual straining prior to release should be avoided.

A clearly understood system of signals should be used between the banksman and the crane driver.

Only one person should be in charge of the operation.

All through pins, formers and stopend connections should be removed prior to lifting.

It should be mentioned that in the case of moulds (particularly battery dividers) the fewer the number of loose parts the less chance there is of loss and damage and delay to cranage. In many cases a careful study of the geometry of a unit will indicate which mould parts may remain fixed and which can be roller-mounted or hinged to reduce the stripping time.

There should be little need for work other than cleaning, oiling and any modification required between uses where moulds are carefully handled. Undue refurbishing detracts from the results of the initial economical manufacture since, minor repairs excluded, the remaking work is unprofitable and excessive rebuilding costs indicate that insufficient attention has been paid at the onset to the requirements of the contract.

It cannot be over-emphasised that apart from pure mechanical considerations, the forms and moulds provided must be substantial enough to withstand the handling involved in casting and striking, and should be arranged to facilitate the placing and vibration of the concrete which they are to retain. Sound construction and care in design will go a long way to ensure satisfactory and economic castings resulting from the mould or system.

Chapter 4

PRELIMINARY CONSIDERATIONS GOVERNING CONSTRUCTIONAL DESIGN

The success of a mould or of a formwork system whether used in the factory or on site depends on the methods adopted being suited to the requirements of the contract in hand. It was emphasised previously that there are countless solutions to any problem regarding the formation of a structure or unit. It is also essential to the success of the work that all trades employed in the various operations can carry on uninterrupted by the other trades concerned.

4.1 INTERDEPENDENCE OF ACTIVITIES

In foundation work the excavator, plumber and drainlayer work on their respective tasks. The plumber may also be concerned with the fixing of a weatherbar or an expansion joint where required by the specification. The maintenance of the formwork and concreting programmes relies on these trades completing their work, especially where cast drainware is incorporated in or under the foundation raft. The operations of drainlaying and back filling must be completed prior to the main structural concreting work. In foundation work and where tank structures are being formed it is necessary to provide blinding before the asphalting is carried out so that the asphalt coats can be laid without causing a delay in the further blinding and screed erection. The bricklayers who provide the skins against which asphalt is laid must also provide suitable areas to allow full advantage to be taken of the quantities of wall forms available. Where special steel linings to openings or cable entries are required for walling this demands careful phasing to ensure continuity of work for all trades concerned. A key factor in such operations will be the early availability of full working drawings with information on openings, fastenings and fixings.

Wall and column kickers which spring from the foundation raft require setting out by the engineer and prompt attention from the steelfixer. This applies also where structural steelwork is being clad in which case wrappings of steel must be available and fixed in place immediately the preceding floor slab has been concreted. Once the concreting has been executed in the walls and columns, carpenters are called upon to erect the formwork for the floor soffits and staircases, thus providing continuous work for the heating engineers, electricians and steelfixers. After this formwork has been levelled by the engineer, openings will be formed and checked, and then concretors can begin to lay the concrete for the floors.

Maximum co-operation between all trades is essential in order to maintain the programme and bring about a successful contract. The carpenter, the scaffolder, the steelfixer and the concretor are required to work closely together on all concreting work, particularly with regard to the formation of staircases and liftwalling on multi-storey structures. It is the progress on these sections of the work that often governs the speed of construction of the whole structure. Stair formation should be completed as soon as possible after the walling operations, since this ensures that the concretor has access without the need to barrow through the flat between the supporting grillage to the soffit supports. It also avoids the need to lower concrete down shafts to the concreting gang. Immediately on completion of the erection of formwork and when the required inspection and checking have been carried out, the steelfixing and concreting operations must follow to ensure that the striking times are kept within those stipulated.

4.2 PRECAST ACTIVITY RELATIONSHIPS

A similar planning of operations is required on precast concrete work. As soon as the carpenter or concretor has laid pallets, the steelfixers or wiring gang must follow immediately with their cages of reinforcement or cable assemblies and helical bindings. The carpenters then proceed with the insertion of formers for the holes and plugs, slots or sockets for subsequent fittings and fixings, after which the side and stopend erection is completed. The moulds are finally prepared by adequate bolting or clamping in time for the concretor to carry out his part of the work within the scheduled part of the working shift.

When the casting has been completed the finishers float off or texture the surfaces as required. After the specified period has elapsed, carpenters or labourers strike out the moulds and in the case of post-tensioned work, the stressing gang render the units ready for removal by the loaders or handling gangs. Finally dressers finish the work in preparation for transport to site.

4.3 PLANNING ARRANGEMENTS

Careful planning ensures that there is a smooth flow of work through all trades, and regular production meetings between agent or manager, and trades foremen and sub-contractor's supervisors, will help to avoid unnecessary delays and that there is correct co-operation between the trades. At such meetings progress to date should be discussed together with future planning, possible modification and alterations to drawings and details. Requests for clarification of existing information can be registered pending the full-scale site meetings between architects, engineers and contractors.

4.4 THE ACTIVITIES OF THE FORMWORK DESIGNER

Where precast or *in situ* concrete work is concerned, a meeting of the supervisors provides an opportunity to examine the proposed formwork or mould methods. In view of the interdependence of the work of all the trades concerned, any resultant discussion of the method employed must prove to be advantageous in the final provision of a system which is suited to the working requirements of those trades. The discussions should establish questions of quality, quantity of formwork to be supplied, requirements regarding scaffold arrangements for erection of forms, and the placing of concrete. Details such as grillages to support projecting steelwork in wall construction and the plant requirements peculiar to the job can be settled. The agent or manager can thus ensure that his purchasing officer is advised of material requirements and the plant department made aware of forthcoming handling and equipment problems. Such meetings help to develop knowledge of the requirements of all the trades involved and the interchange of views prior to the work being carried out avoids delays in its execution.

When the basic outlines of form and mouldwork are under consideration the designer must have regard for the stipulations of the specification which relate to the positions of construction joints, the provision of movement joints and the amount of concrete which is allowed to be cast in each lift or bay. These considerations should be linked to the information that comes from the meetings and discussions between the various trades foremen. Decisions which are derived from such meetings include those of handling methods, sequence of operations, and delivery dates for form and mould material onto the site.

Quantities of formwork or moulds required have to be finalised and the approval of the surveyor or estimator requested with regard to the quantities that must be supplied. The order of priority in carrying out various sections of the work must be established and where precast work is concerned the priority of specific units established. The methods of placing concrete must be related to the geography of the site, the type of work involved and the amount of concrete to be cast per programme unit. Means of retaining moulds and forms must be discussed and all matters which concern the form or mould work and which have a bearing on the quality of the finished structure or unit must be fully discussed with the engineer who is responsible for the work.

When these factors have been thrashed out the designer should satisfy himself that he has a thorough knowledge of all the various items which concern the job. He should now understand the requirements of those who are to produce the finished article and he should have in mind the basic outline of the methods to be adopted. He should now be in a position to design and detail a system for the provision of mould and formwork and draw up a specification of all the constructional methods to be adopted.

From the foregoing it might appear that it is difficult to satisfy the requirements of all parties. In fact all the discussions at this stage of the job usually not only serve to pinpoint one or other of the basic proposals for formwork methods, but also assist the designer with respect to his own decisions on the system which is eventually to be selected. Often, the various parties concerned will have worked together on previous contracts, and their combined experience with previous systems employed provide the basis for assessment of the suitability of any current proposals, and these may well be straightforward developments of those adopted before. Once the designer has formulated an outline of the system his attention will be drawn to considering the materials which are to be adopted in the construction of the component parts of the mould and formwork. His selection is often governed by the economics of the work and the estimator's provisions for form and mouldwork at the tender stage. The choice of material is one of the designer's key decisions, though of course the specification will indicate particular materials which are capable of providing the required qualities of the finished product. Quality of concrete profile and face finish demand that such materials satisfactorily provide the number of uses per form or mould envisaged by the design, and this is a major factor in the selection of materials.

The interaction between material selection and form construction is so great that it is virtually impossible to state which factors govern which aspect. However it is generally the case that the physical details of concrete form, and the quantity of concrete to be cast, dictate the selection of the material. Once the material to be used has been selected the designer can then detail the form construction in such a way as to take advantage of its attributes.

FIG. 4.1. For exposed concrete and board-marked surfaces, panel joints and lift levels must be carefully considered in relation to the lines of the structure.

FIG. 4.2. Pressure measurements being recorded on a test rig for analysis. Formwork research is necessary for obtaining knowledge on such details as the pressure imposed by fresh concrete on a form face.

FIG. 4.3. Large casts and considerable storey heights combine to present major problems for the falsework and formwork designer. Skilful design based on close co-operation is essential to the success of the whole construction process.

FIG. 4.4. The interface between the work of the civil engineer and the plant engineer. It is essential that the formwork designer appreciates the accuracy attainable within each regime.

Chapter 5

BASIS OF MODULAR DESIGN APPLIED TO MOULD AND FORMWORK

5.1 THE OPTIMUM PROFILE

The most vital stage in the design of a mould or formwork is the laying down of the basic profile of the face that forms the concrete. This profile must be the optimum arrangement which suits the majority of the shapes for which the form is required throughout the concreting programme. Obviously the nature of a concrete structure is such that the most efficient method of construction would be the monolithic casting of the complete structure. In practice this monolith is seldom achieved except perhaps where continuously moving formwork is used. Where the scale of the structure does not allow monolithic casting, construction joints must be specified and day joints established. Consultation with the engineer is necessary to ensure that the proposed casting sections are suited to the design of the structure. Construction joint positions are based on the engineering requirements and involve such factors as the amount of concrete that can be placed in a working shift, the speed at which reinforcing steel can be fixed, or structural steelwork erected, and the rate at which associated trades, such as electricians and heating engineers, can install conduit and pipe runs.

Once the sections of the structure have thus been established, they then become the working lifts or separate casting operations of the concrete work. Consideration given to the profiles of these lifts and bays throughout the contract will determine the optimum shape of the individual panels of formwork which must be provided.

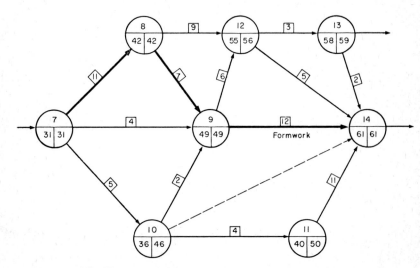

Fig 5.1. This illustrates the importance of formwork operations throughout a contract. Formwork invariably features in the critical series of activities which determine the overall duration.

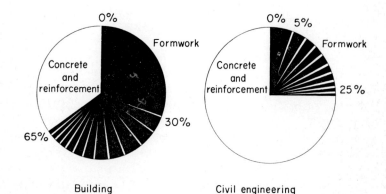

Fig. 5.2. The cost of formwork can be related to the cost of the concrete structure. These pie diagrams show approximate figures for building and civil engineering work.

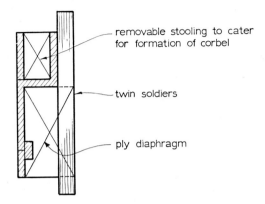

Cater for variations in vertical section at time of manufacture

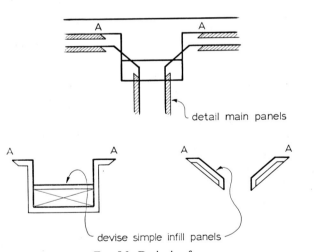

Fig. 5.3. Designing for re-use.

5.2 LIFTS AND BAY ARRANGEMENTS

With *in situ* concrete work, the method of deciding on the optimum profile demands full plan details of the various levels of the structure. Large-scale details produced by the structural or civil engineer are considered on general lines. At this stage details which involve openings in floors and walls are disregarded since later considerations will make provisions for the techniques that are adopted to frame up the formers for these aspects. For walled structures, silos, tanks, multi-storey flats of crosswall construction, lift enclosures in general building works, and chimneys and shafts, the most useful details at this stage will be the sections that are taken through the structure at various positions, both vertical and horizontal. Lines should be superimposed on such drawings to indicate the agreed or specified lift levels, regardless of the floor levels, openings, projecting nibs and similar features except where such features or levels are of a highly repetitive nature and will themselves dictate lift or dayjoint levels.

The lines thus marked on the drawings will become the reference lines for all stages of the work and each should be allocated a reference number for use in succeeding instruction or discussion. It is always easier to refer to a generally known lift number rather than to the level above datum level or local finished floor level. At this stage when the work is complicated in profile and level, simple block models are helpful and often necessary to ensure the practicability of a given casting sequence. The use of models in planning concreting methods is gaining popularity, since the provision of such models simplifies the solution to many questions. Plant location, accessibility of particular units for handling by crane, the geometrical form of the work in electrical plant installations, siphon structures and the like become readily apparent to those concerned with the fabrication and handling of the forms on the actual contract.

With the lift levels established on the vertical section drawings or marked on models, the path of a particular form section throughout the course of the concrete programme becomes apparent. The aim is to minimise the number of handling operations where a particular form has to be diverted from its general course of progression throughout the concrete casting operations. Forms should generally be handled vertically where wall forms are concerned in multi-storey structures, and horizontally and vertically in retaining walls and similar structures. All handling operations outside the general line of travel of the form become expensive. Those concerned with handling are called upon to learn the system and associated handling techniques. Movements and operations which are outside the normal handling scheme and which involve unfamiliar operations can result in wasted time and energy on the part of the operatives that carry out the work. Non-standard operations generally result in temporary arrangements being used for handling. These arrangements may be beyond the scope of the lifting apparatus or travellers employed for the general handling of the formwork and should be avoided wherever possible. In multiple re-use operations, the provision of specially framed forms for use on the irregular sections or non-standard operations in otherwise regular work can reduce exceptional movements.

Where retaining walling, culverts and similar repetitious work is cast in single lifts, and when form movement is generally in a horizontal direction, provision of a traveller is a most efficient means of handling. This applies particularly where cranage is not employed on site or where programmes call for the full employment of the cranes on other parts of the contract. Travellers serve to uncouple form and crane operations and generally facilitate the work of form and transport and handling. For work in which constant vertical movement of the forming system is involved then chain block-lifting or lifting from a centrally positioned hoisting arrangement will prove effective. On building construction work the tower crane offers considerable facility for combined vertical and horizontal handling of the form systems used in walling work, while specially constructed stillages can be used to handle materials used for forming grillages and sheathing to floor slabs and similar work. The table form with its handling arm is a typical travel and handling solution to an operation which would otherwise control the whole rate of progress.

5.3 PANEL DETAILS

Once the course of the individual form components has been established it should be possible by reference to horizontal sections, floor plans and details to ascertain the various profiles which a given form section will be required to cast and so the panel joint positions may be detailed in plan. The panel module is related to both the design module of the structure and the method of handling to be adopted for the component parts of the system. The formation of the structure with regard to breaks at varying levels requires attention as does the alterations in plan profiles. Where fresh walls spring from upper floors, or where sections of the building are carried to higher levels than others as the floor areas diminish, provision is required for adjustment of the outline to suit the layout of structure. Further vertical joints in the form layout will be dictated by the provision for striking of form panels from between return faces, or where reinforcement has to project as starter bars for cross walling which is cast at a later stage.

With the establishment of the plan positions of vertical joints provision can be made for dividing the form system, when required, into sections to be handled by the various members of the labour force as dictated by the priorities of the programme. A series of sketches of the profiles likely to be encountered can now be produced, and an exact figure for the number of uses to which the individual forms will be required can be calculated. The number of uses will dictate the optimum profile to which the basic panels must be constructed.

The basic panel, irrespective of the material in which it is constructed, must be capable of being adapted to the profile from which most castings are to be made. Apart from the foundation work, the majority of the profiles will be related to walling surfaces, which may occasionally have projecting piers or offsets of wing walls and breaks in the building line. This is the profile to which the sheathing of the form panel must conform, but provision must be made for incorporating panels or make ups to form the less recurrent or non-standard parts of the work.

To accommodate projecting steelwork or reinforcing bars, or drastic changes in profile, removable sheeting sections or trimming members should be provided initially for use at an appropriate level. The parts of the formwork system which act as infill panels or make ups should be constructed in a manner consistent with the number of uses which they have to provide. Lighter construction or cheaper material may be used provided that the inserts, over their limited use, form concrete to that part of the specification which governs deflection and quality of the finished face of the concrete.

Wherever infill panels or make ups are inserted, due regard must be given to the basic arrangements needed to prevent grout loss by the butt jointing of components which are suitably constructed to allow ease of striking without damaging the structure. Where only a few uses are to be made of a component it may prove economical to erect a light framing which can extend over several lifts. This framing, suitably sheathed, can remain in place as the main panels are repeatedly erected and struck. The main panel can be eventually struck from the face of the concrete as the standard panels move on. Similarly substantial panels may extend over more than one lift vertically in order to provide a fixed template or support. This helps with the lining of intermediate panels. An excellent illustration of this type of construction is afforded by the vertical members of some proprietary steel systems which remain bolted in place at the concrete face and thus serve to line the panels in succeeding operations.

Where lift shafts, staircase walls and similar box structures extend over many storeys, the provision of panels which form the corner sheathing to several lifts without removal from the face, assist in maintaining vertical accuracy of the structure.

5.4 VERTICAL FORM SECTION

Once the plan profiles of the form panels have been established it is then necessary to examine the vertical sections as dictated by the lift line previously laid down.

Beginning with the kicker, or 'lead up', the main casting heights of the concrete within the form can be ascertained, after which provision can be made for the joints in the form face which make allowance for the projecting steel, or for concrete to be readily placed at a reduced height. The form thus caters for the junction of the wall face and the underside of floor slabs. In these cases it may well prove economical to make certain parts of the sheathing removable in order to ensure correct vibration and placing of the concrete. Where only one face is interrupted by such an obstruction it is always an advantage to lift the outside panel to the correct level for the next established overall lift level. The projecting panel will then create the edge form for the slab and kicker.

Providing that the engineer agrees it is sometimes possible to crank back the starter bars to the floor slab and staircase landings within the face of the main form profile. In this way it is possible to proceed with the main concreting programme and then follow up with the flooring operations at a later stage. Where this is allowed it is extremely helpful to those carrying out the succeeding operations to ensure that battens are secured to the form face at the appropriate level and that the starter bars are cranked to lie between them. The battens remain in place as the forms are struck and can be removed when it is required to expose the steel for projection into the slab. Cutting away of grout in clearing the bars is avoided and if the fillets are sawn to a joggle section, a bearing joint can be formed. This is often considered desirable by some engineers. Where links or binders for slabs or nibs are to be treated in this way, then short blocks sawn to a suitable section and length can be fastened back to the form face within the binder profile. Alternatives to these techniques include the use of foam rubber or plastic for embedding the steel, and the provision of threaded steel and cast in sockets.

When considering the vertical section of a form panel it becomes obvious that it is only the actual panels which encounter a projection or nib which require to be constructed to encompass such a feature. It may be possible to cope with small architectural features within small rebates, or grooves formed in the depth of the sheathing material. Where such features or nibs extend further than the sheathing depth, and often repeat in the path through which the panel travels, stooling members must be provided. The stooling panel should be constructed such that it can withstand the re-uses in which it may be required. These stoolings can be attached to the panel face by nailing or screwing from the back of the panel. They should provide for any reduction or extension in order to locate the feature, as required, relative to the overall lift levels. Care should be taken to avoid the extra depth of the form which can spoil the striking arrangements. Stoolings may call for the insertion of extra striking fillets to allow withdrawal from between reveals.

5.5 CONTINUITY AND SUPPORT

It is necessary to consider the economics regarding materials and labour which are involved in packing out an existing panel to form a profile. In some circumstances it may be cheaper to provide a special section of formwork to overcome the problems of casting large projections and nibs that are constantly repeated. It is certainly uneconomical to pack out any great part of a panel area in order to form a shallow projection at the top of the lift, although sometimes this may be necessary to maintain a given work programme. For example, the expenditure required to pack out or stool for a projection which occurs in only one face of a silo enclosure may well be repayed by the continuity of a casting which is maintained over the remainder of the walling in the same lift level. Splayed or battered walling can easily be dealt with in square or rectangular structures by the use of varying widths of top spacers and lengths of ties, to provide the required profile; tapered make ups can be added at corner panels. The depth of lap back over previously cast concrete must be regulated as must the position of the tie or anchor to avoid springing of the form face or irregular profiles. Where a modification is required to the sheathing profile of a circular form, the cost of adjustment may well prove excessive. As soon as a panel becomes displaced from its designed position the radius will be incorrect, the chord length will cease to match its planned dimension and any adjustment constitutes almost a re-build. Where extremely large radii are employed the facet formed by a panel becomes exceptional, and is quite marked, although where commercial concrete is provided the deviation from the specified line may be acceptable. On small diameter silos and bins any movement from the planned position can drastically affect the concrete profile and will demand an expensive form adjustment.

Hoppers, be they square or conical shaped, call for considerable form adjustment for a particular variation in pitch, or variation in concrete thickness, and each form component must be considered with the possibility that it may have to be replaced or adjusted. An optimum profile should be established whatever the module size selected, or profile of the individual component governed by the concrete. This profile can be modified by stooling and packing to cater for any exception arising or necessary modification.

This chapter has been concerned with both the plan and sectional profile, and the means of adjustment to these features. The next step in form design is that of ensuring that while the sheathing is capable of generating the required profile, a suitable backing system or grillage must be designed to collect and transfer the loading into the tie or supporting system.

An essential requirement for successful form design is the

introduction of members which ensure the continuity of the system. The line of the form face must be maintained and differential movement between the form panels and components must be avoided. Soldiers, walings, joists and bearers in traditional systems and strongbacks, soldiers and centres in proprietary systems provide the means to meeting these requirements.

The interrelations between the tie arrangements and the supporting system are considerable and demand careful study. A major part of the economics of a system hinges on achieving a good balance between tie positions, sheathing deflections and economic support arrangements.

The positioning of tie rods, ties or anchors is an essential part of formwork design and every advantage should be taken of the opportunities that arise for supporting secondary grillages or form arrangements from the anchorages normally cast to retain formwork in the principal operations. Although it may increase the formwork cost to leave extra ties or anchors over those required in the primary set up, with proper care in the planning stage the ties can be arranged to suit the purposes of multiple operations and thus the maximum value can be obtained from the tie system.

Although the selection of ties, bolts and similar ancillary equipment is considered later in this book, it must be stated here that the designer should consider the tie system on the basis of the specification which governs the concrete structure, the simplicity of the tie arrangement and the versatility of the tie. No one system is universal and while snap-ties can prove to be particularly effective in building work, a similar piece of concrete work on a civil engineering contract may well call for a completely different tie system. Although sustaining and transmitting loads are vital requirements, the material and configuration of the tie plays major parts in form and mould design, sizes of components and sections of members.

5.6 PRECAST CONCRETE—MOULD DESIGN

Considerations with regard to modular design as applied to precast and prestressed work follow the same pattern as those for *in situ* work. Precast work does not generally lend itself to any division into sections for casting purposes. In the majority of instances the concrete for precast units is cast monolithically. For precast work the mould panels have to fit together to form units of concrete which in their turn combine during the erection sequence to form a complete structure. Every precaution must be taken to avoid dimensional error and consequent difficulties in the erection of all structural work. Where the structural engineer works in close co-operation with the manufacturer and pays due regard to the practices generally employed in the casting operations little difficulty should be encountered. As many of the precast components are produced in works which are situated far from the sites on which they are to be erected, close liaison between the engineer, the precast works operatives and the site personnel must be maintained to ensure that manufacture follows the programme, that delivery to site is as required, and that accuracy of units, which allows speedy erection, is ensured.

Delays in casting, while affecting the output of the factory, are further reflected in delays in the erection process, and may result in extra cost of plant and the provision of additional equipment. Contractors frequently rely on having to hire heavy lifting plant, and charges on such machines are usually too high to allow demurrage through production difficulties. Early receipt of finalised details in the works help to eliminate delays. Where precast products are manufactured for various concerns there is a great need for the dovetailing of contracts to ensure a smooth flow of production, and every bit of information is vital for efficient production control. Transport from the works to the site requires careful organisation where large loads are involved. Route arrangements and permission must be obtained from the various authorities for the transport of abnormal loads. Special notices thus render 'spot' deliveries impossible.

A few years ago, the greater part of precast concrete work would have involved the production of monolithic units, but now the trends in prestressing together with the greater demand for large span bridge beams have tended towards the production of units cast in sections, and subsequently post-tensioned into the final form. Recent major civil engineering works have included a variety of forms of segmental construction. Precast work is often carried out under factory conditions free from the delays that can be caused by the weather. A greater degree of quality control and intensive production facilities, coupled with improved transport and cranage facilities, are now leading engineers and contractors towards precast units which would, some years back, have been considered essentially works of *in situ* concrete.

In an effort to reduce on-site time and supervisory commitment, contractors are precasting items such as ducts, subways and even retaining wall sections for later incorporation on site by post-tensioning or linking by small infill sections of *in situ* concrete work. Whatever the product, the principles of modular design of the mould component apply. As with *in situ* work there is an urgent need for finalised details of the units to enable a comparison to be made of the types of unit and overall planning of the provision of moulds. The engineer's details and specification are required to establish profile, all quantities and details of finishes and a further consideration which is not encountered on *in situ* work, that of the mass of units. The vital links between the precast works, or even between the site precast bed and the erection site, are transport and cranage and the question of the handling of the finished product figures prominently in precast and prestressed practice. The pile yard will have either a derrick or gantry crane; the precast yard, similar cranes, or perhaps mobile cranes, while the works may use travelling cranes. Castings need to be limited to such weights as these plants can handle. Moulds for heavy items of precast work can be constructed in large sections since plant is unlikely to be continually employed on lifting operations as is found on site, except where overhead cranes are used for placing concrete.

With the drawings to hand and a delivery programme laid down, the required units can be divided into groups of a similar nature. Piles of similar section, beams and columns of like detail can be divided into programmed production units, and the decisions as to quantity and quality of moulds to be provided can be assessed. Where work has been specifically designed for precasting, this will obviously present little difficulty. Considerable economies can be effected by engineers in the simplification of sections and details, particularly with regard to the connections between units.

5.7 THE BASIC PROFILE

With the groups of units isolated, the profile of the basic mould can be determined and here regard must be given to the numbers of units specified for a given section and the outline to be cast in the overall programme. It is possible with present day materials to construct a mould capable of providing a high re-use value. The components of such a mould should be so constructed as to form a plant item for possible use on a variety of types of product within a works. Care in standardising design will do much to reduce mould costs and cut down on the high cost of providing moulds. 'One off' units called for in some types of work such as corner columns can be manufactured more economically when the moulds have been carefully devised for re-use.

The basic profile selected will be such that it will encompass the largest part of the eccentricities of units in a given construction batch. Production will require to be organised in such a way that particularly irregular units may be cast after the more typical units. In this way any modification to the mould equipment will not affect the quality of finish or delay the production of the more standard items. It is a simpler matter to govern the basic mould profiles with precast units than with *in situ* concrete as work may frequently be executed in a sequence to suit the mouldwork. In cases where particular unit details may over-complicate the basic profile, the engineer's assistance may well be sought in allowing the precasting of nibs and features. These components are then available for incorporation into the unit at the time of main casting. In heavy operations the unit may be cast with any one face on the mould bed and careful adjustment of the way of casting can cope with any particularly awkward projections. This aspect or *way up* of casting is a critical factor in mould design and is dealt with in depth elsewhere.

The profile which embraces the majority of the units to be cast in a given group decides the eventual line of the sheathing face of the mould. Pads and packers are inserted within this profile to deal with eccentricities. Length variation is the simplest variation to incorporate within a mould since stopends can be easily re-positioned and cleated into place.

Variations in the top face of the unit as cast can readily be formed by small upstand formers which are planted on to the upper edges of the sheathing. Bearers or cleats should be made sufficiently long in the first construction to provide support and continuity of face.

Variation in the base of the unit as cast demands forethought in the formation of pockets and recesses in the pallet with suitable fillers and packers to ensure ease of casting of the more standard units, especially where the mould is employed to cast such units with plain faces or perhaps smaller nibs.

Nibs and projections on the sides of the unit as cast can be incorporated within the pockets which are themselves arranged within the side sheathing, while large variations can be overcome by the replacement panels inserted in the side panel arrangements. Here variation of infill panels allows casting of various levels of crane support nibs or wall panel supports.

Where large apertures are to be cast through the unit to accommodate services, they are best cast through the vertical axis of the unit as cast on the bed. This facilitates concrete placing, otherwise difficulties can be experienced in ensuring the correct consolidation of the concrete below the horizontal faces of the aperture formers. Longitudinal ducts of large section or formers to hollow units present a similar problem, and call for the provision of removable sections of side forms or apertures to provide access for the insertion of poker vibrators. Careful placing of through dowels or location by blocking pieces from tie rods is necessary to prevent float of the core pieces.

The positions of through-holes and ducts and projecting steel must be plotted on the basic profile to ensure that the sheathing bearers are placed in such a fashion that they will not coincide with dowel bars or bolts used to retain bearing plates and duct formers. Having established the sheathing profile and bearer positions, the detail work can be started and drawings produced for mould fabrication purposes.

Where prestressed concrete work is concerned the same procedure regarding mould outline must be observed. Additional steps in the design include: the detailing of profiled panels for forming anchorage blocks and means of positioning ducts and tendon grillages. Details around cable anchorage positions need to be carefully studied to ensure that accurate seatings for anchor plates and cones are provided at the correct angle of incidence with the unit ends, thus ensuring flush seatings and equal distributions of load from the tendon to the concrete unit. Where individual concrete units are to be post-tensioned into complete large-span units, or where precast diaphragms are to be linked by *in situ* concrete into hollow beams, precautions must be taken to provide extremely accurate through ducts for the stressing cables. Stopends to individual blocks must be carefully arranged with regard to the width of dry pack spaces to form continuity of the specified duct profile. Where close tolerances are specified to govern the profile of the units, it will prove to be most effective if the mould is manufactured slightly smaller than the maximum dimensional tolerance allows, since inevitably there will be spreading of the mould when the concrete is placed and it ensures the best use of the permissible deviation will be made in this fashion. Where artificial stone facing is demanded and units are to be dressed mechanically by grinding, moulds should be constructed slightly oversize in areas where such a finish is specified. This allows for the eventual dressing away of the stone particles of the surface.

The foregoing description of the basic method of designing form and mould systems to ensure enhanced re-use value can best be applied where manufacture is carried out prior to erection. The application of the technique is essential in cases where forms are constructed in contractors' workshops or in the factory of a specialist supplier. The basic panels, limited to the minimum number of types, must be substantially built to ensure an economic formwork system which provides the uses required with the minimum refurbishment being required while the panels are in use. Site alterations and reconstruction of form panels are expensive and may well result in delays in the casting programme. Once the basic profiles have been established, the methods of construction can be established. Such questions as the amount of lap over previous castings, means of tying adjacent faces and the outline of the supporting system of steel or timber members can be decided. In walls, beams and precast work, twin soldiers are frequently used to facilitate insertion of tie rods, bolts or patent ties and these are required to be positioned accurately on the respective panels. Wherever possible ties should be placed normal to the sheathing for which they provide support. The means of handling form panels will dictate an appropriate provision for lifting.

The mechanical properties of materials govern the spacing and section of the supporting components. Simple calculations must be made to decide on the standard sections and spacings that are to be applied throughout the system. Following such calculations comes the preparation of the general assembly drawings and then the details for individual panel construction.

The value of a general assembly drawing, or drawings, which indicate the exact location of panels when at work cannot be overemphasised. A clearly presented drawing simplifies the erection process, eliminates wrong placing and unnecessary adjustment of panels as well as simplifying communications.

Chapter 6

DETAILING THE FORMWORK SYSTEM

6.1 CHOICE OF MATERIAL

Once the mould and form profiles have been established by the means outlined in the previous chapter consideration can be given to the choice of material and the methods of construction to be employed. The profile goes far to govern the broad choice of materials to be used. Where there are large numbers of castings to be obtained from a panel with little or no alteration to be made to the profile, then it may prove economical to use either purpose-made steel forms or one of the proprietary steel formwork systems. Where there are constant alterations or a low number of uses of similar profile it will certainly prove economical to use timber or timber derived materials. This applies equally to *in situ* and precast work, often the smaller mould items in each case may be cast using plastic or reinforced resin moulds.

Large units can offer great scope for use of concrete both for pallets, and with some co-operation between manufacturer and engineer, side forms and stopend units. Combinations of materials can be used to gain the advantages resulting from the diverse materials available. Steel walings or centres may be used with timber of ply sheathing and timber features may be fixed to a steel sheathing for precast and prestressed work.

6.2 PROVISION OF DRAWINGS AND DETAILS

Once a decision has been reached as to the materials to be adopted it becomes necessary to convey to the manufacturer the profile and methods of construction to be followed, the same basic information being required by the steel fabrication site carpenter or contractor's joinery works staff.

Although there may be accepted constructional practices in existence within a firm, just as there may be standardisation of

Fig. 6.1. Casting operations were continued under rigorous conditions when these steel forms were electrically heated.

FIG. 6.2. Telescopic centres provide a simple means of sheathing support. However, provision must be incorporated for removal from beneath the newly cast concrete.

sectional sizes of material for given application, it is essential that there should be drawings and a specification, however brief, for items of mouldwork and formwork constructed by one department or concern and used by another.

Where the mould and form design is carried out by a specialist consultant, drawings are prepared by the consultant's office staff which show the general arrangement, detailed lift levels and schedules of panels fully dimensioned, complete with lists of ancillary steel work and ironmongery. Such lists can be utilised in all grades of manufacture whether in the steel shop, site carpenter's shop, contractor's joinery works or subcontractor's works. At the other end of the scale sketches and details, often worked out on site and in mould shops on scraps of paper or timber can still result in the provision of satisfactory moulds and formwork. It is advisable always to work to a standard procedure in preparing mould or form details for presentation to those responsible for the manufacture. Only in this way can errors be minimised and availability of parts be ensured as and when required in the works or on the site. The main steps in establishing the profiles of the panels or moulds have already been discussed in the previous chapter. Following the choice of material based on the desirable properties, the following require attention whatever type of material is selected.

6.3 GENERAL ASSEMBLY DRAWINGS

General assembly drawings must show the complete plan of the structure at the main lift levels, where the majority of uses with form panels set to a given profile, are obtained. The positions of all tie rods and bolts must be indicated as must panel joints, striking fillets and lifting point arrangements for crane operations. It is also useful to line-in scaffold arrangements, hoist and mixer positions, placer or pump positions. Each individual form panel unit should be allocated an identification number which is marked indelibly on the appropriate unit at the time of manufacture. This general assembly drawing will be constantly used in the detailing of the individual form panels. Such details as the facility for striking between cross walls, placing of buried anchorages for succeeding work, crane radii for mechanical lifting and block positions for manual hoisting will readily become apparent. The general assembly drawing brings to light such items as the particular positions where through bolting may become impossible due to buttress walls or fillets rendering the wall section too deep to allow bolt withdrawal. The position of architectural feature can be related to panel size, while panel joints are dictated by the main changes of section apparent when assembly drawings of the main plan arrangements are compared. If the structure is of a particularly repetitive nature as regards the profile of lifts of concrete, the general assembly drawing with its clear, sectional details of panel treatment may well be the only item of drawing work required to enable the manufacture of form panels. Vent shafts, ducting, runs of retaining walling and similar work can be detailed in this fashion and the drawing handed immediately to the manufacturer.

6.4 SECTIONAL DETAILS

The next operation in the case of a more complicated concrete structure is the production of details showing the various forms in their casting positions in the non-standard levels of the job. A study must be made of positions where intersections of walling with floor slabs occur and where projecting steel or structural concrete nibs are to be accommodated. The formation of major openings in walling which extends over several lifts must be indicated and a sectional drawing can be used to indicate the positioning of panels which may

Fig. 6.3. Models can be of considerable assistance in the planning and detailing of formwork operations. This model is used for training purposes.

Fig. 6.4. Excessive overlap or clip back on to previous lifts of concrete can produce problems such as 'curtains' and honeycombing—evidenced at this horizontal joint. Where appearance is a critical factor, foam sealing strip should be used.

Detailing the Formwork System 63

remain in position over a number of lifts in order to provide continuity and add to the rigidity of the system.

The sectional drawings which relate to a formwork system should have lift levels clearly numbered with heights above datum shown,

formwork and moulds must be designed to strip out from within returns or re-entrants

smooth faces and pencil rounds assist in all striking and stripping operations

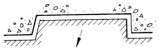

a lead or draw on one face is sufficient to allow withdrawal

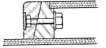

where no lead is allowed artificial lead can be provided by stripping fillet removed later

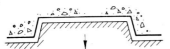

a lead or draw on both faces is better

alternatively lead can be built in by splay cutting form panel rendering section 1 removable first

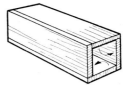

boxes of correct construction can be folded out

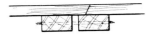

panel joints can incorporate lead for striking

FIG. 6.5. Lead and draw.

and must indicate all through bolt and tie positions both for form panel retention and for fixing formwork and support for subsequent operations. This detail regarding tie points is important, especially where either buried tie or through bolt systems are being used to ensure that suitable anchorages and through-holes are available for the subsequent erection of forms for split lifts or lifts incorporating floor edging. Care in positioning ties can reduce the number of drillings made through the sheathing at differing levels for tie rods and bolts.

6.5 PANEL DETAILS

As soon as the general assembly drawings and sectional drawings have been completed, details of individual panels can be prepared for the manufacturer's use. The basic sheathing material will have been selected, and from the sectional drawings the deepest lift can be measured to allow calculation of the sheathing thickness, backing spaces and section, waling position and section and tie sizes. It is advisable to standardise these sizes and sections to an overall economical arrangement which is capable of carrying out the more typical operations. After adequate calculations have been carried out, accurate minimum sectional sizes can be laid down for each member. For particular applications however it will often prove to be of greater advantage to utilise larger, and thus lightly stressed, standard sections of material which can be bought on to site from standard sections of material which can be brought on to site from machining often can be eliminated. The majority of proprietary steel systems work on this basis by the provision of a set range of supporting grillages. Often consultants and specialists adopt the procedure for run-of-the-mill work. Calculations are generally only involved for special applications where such standard formwork units cannot be utilised. Peculiarities of profile, extreme depths of concreting lifts exceptional thicknesses of slab and inaccessibility for propping and shoring introduce a need for calculation and reference to standards.

With the establishment of the sheathing thickness and section and spacing of grillage members, the individual panel details can be prepared to a large scale. Overall sheathing areas must be indicated, as must the position of cleats and bearers and their sectional sizes. The dimensions which relate to bearer and soldier positions must be accurately given to ensure lining of bearers for through-tie insertion.

Running dimensions from prominent features or returns indicated on the general assembly drawing can readily be associated with panel joint construction to ensure accuracy of fit. Bearer positions in the horizontal plane should be related to the end joints of the panels or specific points on the structure in all cases in order to facilitate subsequent setting out during the construction and erection processes.

The plan and elevation of an individual panel should be accompanied by sufficient sections to indicate the designer's requirement as regards boring for through-ties, blockings between twin soldiers and borings for connection to adjacent panels in the same lift. These sections also serve to show provisions for the formation of split lifts such as sheathing joints and fillets which form ruled joints at the top of lifts. Rider drawings should be prepared to show pads and fillets for use with specific panels in the formation of projecting nibs and features, together with the method of fastening to the main panel duly indicated. Where timber or tube spacers are used, in conjunction with tie rods, typical details assist in the manufacturing process provided that quantities and sizes are stated. The use of buried ties combines spacing with the tie method, and when scheduling ironmongery requirements the detailer must note the appropriate size of tie and cone as well as giving an indication of the capacity required.

Most of the constructional details for geometrical work can be conveyed on the general assembly drawing, and since such work frequently calls for full size setting out in works, including the preparation of templates, there is little to be gained from drawing individual parts of the form to scale. In this case however it is essential that the general assembly drawings should convey all the key dimensions for manufacture. It should be possible in all cases to produce the moulds and formwork from the formwork details alone without further reference to the original drawings produced by the engineer or any other source. Limiting the number of drawings to which reference has to be made reduces the possibilities of error arising from modification or amendment of drawings and introduces a discipline on the detailer to ensure that complete information is available.

6.6 BAY LAYOUTS AND BEAM DETAILS

The foregoing description of form detailing has revolved mainly around the provision of walling panels and unit moulds. The provision of general assembly drawings and details are equally important for floor bay sheathing and beam forms. The planning of form panel layouts for floor sheathing can result in a great saving in the labour and material expended on the insertion on site of make-ups and rips which assist in lining out irregular areas with the modular panels of a chosen proprietary system. Apart from providing an instructive guide for faster sheathing, the layout ensures that joists and bearers are used in such a way as to minimise the amount of end cutting with consequent wastage of material. Careful insertion of rips or continuous make-ups ensure that joists are laid to pass over a bearer and oversail to avoid the need for trimming. A drawing of the plan arrangement of sheathing units is essential where drop panels occur around column heads. A layout for slab soffit sheathing where there are numerous secondary beams spanning between main beams is likewise essential. Another case occurs where column dimensions are greater than those of the main beams or secondary beams. Here the small areas of sheathing adjacent to the column corners require care to ensure that adequate support is provided for make-ups and fillets.

An accurate drawing of the joist and bearer arrangement related to a site feature, a line of columns or walling enables a rapid positioning of the telescopic props, exactly where they are required for the support of the bearers to the grillage system. The formal location of the prop positions and centres is essential for safe working and is the only means of ensuring correct prop spacing in the absence of proprietary tie bars or spacers between props. Where joists and bearers are marked or branded with identification numbers they can be sorted out after striking, and handled in batches from bay to bay or floor to floor, and can be selected for use as required.

Layout drawings for beam forms are used to indicate the position of the form panels while in use. They must be related to the joist and panel layout for the soffit sheathing. According to the system chosen one item obtains support from the other and thus it is important that the correct sequence of erection and striking is indicated. Where there are many uses of beam sides, with little alteration to its construction, it is economical to incorporate whatever make-up timbers are required within the framing of the side forms, and thus reduce the amount of items to be handled from floor to floor.

With regard to the scale of general assembly drawings which relate to mould and formwork, it will prove to be most useful to those who read the drawings if the details are laid out to a scale of 1:10 or 1:20. These scales allow sufficient detail to be included where the construction is of a simple nature, though of course full-size sections and sketch views of details greatly assist in the interpretation of the drawing.

6.7 SKETCHES AND SCHEDULES

Sketches are particularly useful as it is often extremely difficult on the more complicated plane and geometrical work to indicate intersection lines and re-entrant faces, or features on a simple line drawing. Cut-away sketches of a form or mould which include the mark numbers of the required panels can be of assistance to the assembly gang in the first uses and can also indicate the sequence of striking. Schedules of panels, ironmongery and special fittings required for use in casting should accompany the general assembly drawing and detail drawings. These can be conveniently scheduled in the order of the panel mark numbers as detailed on the general assembly drawing. Particular attention should be given to the 'handing' of form panels on such work as silos and storage bins where large numbers of panels are employed that are almost identical but which vary perhaps by one bearer, or by some detail being handed to another due to a slight dimensional difference.

When the details and schedule are being prepared it is advisable that the detailer should have a copy of the general assembly on the drawing board so that the panels can be marked off as detailed and scheduled. The panels can be numbered sequentially, the letters being added behind the mark number to denote associated components, pads, stools, fillets and such like. Figure 5.3 indicates the type of drawings which can convey the proper information from the designer to the manufacturer.

When such details are passed from one firm to another, there should be a descriptive order and specification which govern the manufacture and materials to be used in conjunction with these drawings.

It is necessary on all work which proceeds in lifts with great variation in the arrangement of the form panels to detail the nonstandard uses, and to ensure that the pads and stools required for the formation of the profile are sufficiently detailed to enable their manufacture or assembly within the form as required. Care must be taken over such items as obtaining the correct length of through bolts, tie rods and hangers, and bolts for use on the buried anchorage system. While blocks can be inserted to absorb the unused portion of a tie rod or bolt, this is a practice which is both dangerous and hazardous. If the blocking pieces have been on site or in the works for a considerable time they may be over-dry, in which case they will crack and drop out, and thus allows the form panel to move out of position. Alternatively, wet packings may compress during concreting as pressure is brought to bear on them and the water is squeezed out with resultant misalignment of the face. Stock which has been carefully ordered will ensure suitable sized ties are available on demand.

6.8 MOULD DETAILS

Where a precast or prestressed manufacturer's mouldshop is of sufficient size, the production of mould details, similar to those described for *in situ* formwork, will be required. It is a general practice with the smaller items of mouldwork to issue the mouldmaker with a copy of the engineer's unit drawing. These drawings should be accompanied by sketches which indicate the way up of casting, and face of unit to be incorporated into the pallet. The tradesman produces, based on this information, a mould on the lines of the accepted methods of mould construction employed within his works. While this method proves satisfactory with moulds produced by mouldmakers working alone or in pairs, immediately more than one pair of carpenters are required to construct a mould, either through reasons of programme or size, mould drawings should be introduced as a means of conveying instruction. This ensures that component parts match, and together form a satisfactory mould on completion.

Many precast and prestressed concrete producers maintain a mouldshop which is capable of manufacturing moulds for an average production programme, and which employs the services of a subcontractor to produce moulds for unavoidable peaks of production. This calls for the provision of mould drawings and a specification for the sub-contractor to ensure that 'bought-in' moulds are suited to the concreting method and are constructed in line with the general practices of the concrete manufacturer. Operatives trained in a particular method of working with regard to the assembly of moulds only assimilate changes in method slowly, however slight the variation demanded by the way of working.

In many respects mould details are similar to the details produced for individual formwork panels for *in situ* concrete work with perhaps the difference lying in the fact that one drawing will generally suffice to convey both the general assembly arrangement and the panel detail. The mouldmaker will find it of great assistance if the outline of the unit to be cast is indicated, and the various positions of nibs and features which require to be incorporated in subsequent casting operations are superimposed on this drawing. A carefully produced mould drawing can incorporate sufficient detail to enable the mould to be completely set up without further reference to the engineer's unit drawing. The drawing should, where possible, include modifications to cope with awkward situations, for instance where steel reinforcement protrudes from a concrete face which demands alterations to slots and openings provided in the initial set up.

Where there is a factory inspection department responsible for the finished concrete product, a dimensional inspection of the mould against that shown on the engineer's unit drawings will provide a two-fold check on the product of the mouldshop and the mould designer's detail. Much of the planning of the sections in the detailing of precast and prestressed work is concerned with problems of the provision of striking fillets, leads and those component parts which are capable of resisting the increased infiltration of grout brought about by vibration applied within mould sections. Moulds are also required to resist the considerable rigours induced by the use of external vibrators strapped to them. Generally, mouldshop machinery is capable of producing the accurate sections of material required to form complicated profiles. With the close tolerances that are needed to govern such items as anchorage positions for stressing cables and bearing plates, the requirements of the mechanical design are frequently superseded by those of the provision of components which have to be constructed to resist wear, scour and handling, rather than to meet the strength requirements necessary for retaining the fresh concrete within the mould. The details should show the section and elevation of the assembled mould together with notes regarding inserts to be attached to the mould face in the initial set up. Where a mould is to be modified by the insertion of blocks or nibs in subsequent castings, provision for their inclusion and fixing must be clearly indicated.

Individual details are required for items such as pallets, side members, spacers and stopends. A drawing is necessary for showing the number of concrete pallets required or glass-reinforced plastic mould components to be manufactured by other departments. Where two-stage casting of the product due to irregularity of profile is used, care must be taken at the detailing stage to ensure that the positions of through-holes and anchorages, in the first operation, are such that they will be suitable for re-use in the second casting arrangement. Where the casting deck of the works floor consists of a rolled steel joist grillage, or possibly precast sleeper foundation, the base bearers must be positioned to bear on this grillage. Where nibs or corbels are likely to project beyond the pallet structure, instructions on any stooling or extra sleeper foundation work necessary must be incorporated on the drawing.

All intelligent detailing for form and mould construction demands an intimate knowledge of the steelfixing method to be adopted and the techniques of the placing and vibration of concrete. It is essential at the detailing stage to remember these methods and to ensure that the mould constructional methods are suited to the material that has been selected for the construction form and mould. The designer must make every attempt to incorporate the provisions discussed during the initial planning stage between the representatives of the various trades concerned, and he must also remember the plant arrangements that are available on site or in the workshop. Small details of mould or form design can be easily built into a system at the drawing stage, thus saving time and labour both at the mould or formwork construction stage, and at the production stage on site or in the works. If the design is essentially practical, and based on simple principles, it must succeed.

Dimensions indicated on smaller mould drawings are shown as individual dimensions, but with moulds and panels which form part of long-line moulds, it is necessary then to employ running dimensions. Mould details must indicate not only the materials to be employed, but also show clearly where components are to be rigidly fixed by welding or glueing, and where provision has to be made for the removal of features to adjust profiles. The details must indicate the means for accommodating varying lengths of features such as are found with I-beams and those situations where, for instance, diaphragms for the transverse stressing cable interrupt the overall run of the feature on the intermediate beams and which have been omitted on parapet beams. Often in precast work a mould panel need only provide a rigid, partially sheathed grillage to carry profiled pads or formers that cast varying profiles. Standardisation of components thus ensures re-use of a variety with only the minimum of adjustment and refurbishment being required. In this way it is relatively simple to render the moulds serviceable throughout a range of operations, and thus provide an economical form and mould arrangement.

An example of this latter approach occurs where tilting frames or tables provide the rigid carcass to a mould or where modifications required in succeeding contracts are achieved by re-lining or by the insertion of stools or packers.

In a further precasting technique, battery moulds provide the overall profile of a particular product, wall or floor unit, and stools, packers, spacers and stopends are used to modify these profiles for succeeding contracts.

Chapter 7

FORM AND MOULD MATERIALS 1—PURPOSE MADE AND PROPRIETARY STEEL FORMWORK

The materials available to the designer for specification in mould and form construction can be divided into three main categories, namely steel, timber and timber derived materials and modern materials such as concrete and plastics.

7.1 STEEL FORMWORK

Steel formwork arrangements can be of two kinds. Those that are purpose made with a particular contract in mind or those which employ adaptable systems of proprietary manufacture. A well designed purpose made steel form or mould provides the most reliable and economical unit or system of formwork. The cost of such a form is such that it can be used only when there are considerable quantities of a given section of precast work or several hundred yards of *in situ* work of similar profile. The cost in this instance can be reduced to such a reasonable unit price that it compares favourably with those which include the provision of shorter lived materials to form the work. These particular forms and moulds may call for refurbishment or replacement during the course of a contract. There can be no doubt that the purpose made steel form which often includes an integrally built traveller is the most efficient arrangement that can be provided to form *in situ* concrete.

The purpose made steel form is usually the result of the combined efforts of a design team of engineers who have worked together over a period of some months. It comprises a mechanically sound and constructionally correct arrangement of formwork that is capable of being erected and struck with the minimum amount of site labour. The arrangement provides good quality face finishes such that only cleaning and oiling is required between uses. Frequently when purpose made steel forms are used, the services of qualified erectors who are capable of training the contractor's site personnel in the use of the system are called upon. At the same time there will also be a back up of technical assistance in the event of difficulties arising from the use of the system. One chapter in this book has been devoted to the philosophy and manufacture of the larger purpose made steel forms, since the fabrication of steel forms by welding and pressing is the province of the engineer.

In the precast products field, hard landscape items, pile casting and the formation of standard section beams and flooring units present extremely economical uses for simple purpose made systems and moulds. This also applies to the production of small cast concrete units such as posts, brackets and unit system structures. The majority of such forms consist of heavy sheet steel on framing and of simple profile. The main reason for the selection of steel as the constructional material lies in its durability and resistance to the handling which results from the rapid processes of casting and striking. Smaller purpose made steel forms are of sheet steel sheathing supported on a channel or angle steel backing which has the advantage of slight flexibility which assists with the striking operations. The hard smooth texture of the face of these forms imparts glasslike finishes to the completed concrete product.

The second type of steel form and mould construction is that in which proprietary modular systems of formwork are available on a hire or sale basis. Here again the support of the expert designer is made constantly available. A design team has studied the design of the component parts, development has been carried out over many years in order to provide fully proven modular systems which are capable of being adapted to form most of the profiles that are encountered in the casting of *in situ* concrete work. The product generally takes the form of a patent modular panel which has a supporting system of adjustable props or telescopic centres. Recent developments include special ties and anchors to retain a variety of formwork systems during the casting of the concrete and also lattice-type girder units which offer large span supports to sheathing systems.

Manufacturers have also turned their attention to the provision of timber or ply-faced steel framed panels for wall and slab construction, these offering great speed of erection and economies in labour. As there is such a wealth of informative literature available on request from the suppliers of proprietary steel formwork systems, it is not necessary to describe the various components to any great length though it may prove useful to discuss the selection of a system for use on a site or in a works.

7.2 SHEATHING ARRANGEMENTS

Mention has been made in an earlier chapter of the advantages to be gained from the standardisation of one particular system throughout a firm or works. The smallest difference between the finished thicknesses of panels, or the means of forming a connection between adjacent panels, can render the panels unsuitable for use on the same section of sheathing. Most systems depend on the attachment to the walling or bearer members by special patented clips for rigidity but where it is not possible to insert these due to the use of an assortment of makes of panel, the panels could be only useful as sheathing for slab soffits. Out of the systems available it may prove economical to select the most suitable components for the type of work generally carried out by the contractor, the panels or plates from one system, the props and telescopic centres of another and so on.

As there can never really be a suitable solution to a given formwork problem, there is no perfect proprietary system which meets the requirements of all types of work. Although manufacturers reach near perfection for most types of work the universal solution for all applications has yet to be found. Manufacturers freely give advice on the suitability of their system for a particular item of work.

In the first instance it is necessary to decide on the type of sheathing panel to be hired or bought in, and it is advantageous to accept that a modular system can work satisfactorily only to within 100 mm or so of the required bay size in soffit-sheathing work. Considerable effort can be expended in endeavouring to fit modular units into irregular bays between beams, whereas frequently the work could be performed more expediently by sheeting out the basic shape to the nearest panel size and making up with timber fillets and boards to the exact bay size required. Some proprietary systems offer overlap panels to enable irregular panel layouts to be formed, and these offer an improvement on attempts to use the smaller sizes of modular panels in make-ups. Delays that are likely to be caused by searching for particular sizes of panels for use in make-ups are thus eliminated. The insertion of the extra joists that are required at panel joints can be avoided by the use of single joist members adjacent to the walling or brickwork, the infill arrangement can effectively span between this plate, or beam top member, and the face of the nearest standard sheathing panel. With the systems that employ ply-faced steel frame panels, provision should be built in for the support of ply make-ups to overcome variations beyond the standard module.

The type of panel adopted is closely tied to the specifications which govern the required finishes. The most satisfactory panel with regard to finish is the sheet steel panel or film-faced ply unit which has an angle steel backing that can be laid close to prevent grout loss. The use of these panels is generally accepted as providing the finish to wrot formwork standards.

For work which will subsequently receive some applied surface finish, a simple steel panel will be sufficient, although care must be taken to remove the fins or nibs which occur at the joints from the face immediately after striking.

As it is necessary on some contracts to pass sheathing units from floor to floor by hand, a form panel should be selected which combines rigidity with lightness. These panels are used with the simple snaptie arrangements, requiring only the addition of tubular walling and bracing to provide a complete form. A whole new field has opened up where glass reinforced plastics and in some instances steel, concrete or paper waffle formers have been adopted in formwork. The resulting structural floor is mechanically an improvement on previous structural floor arrangements for large spans and thus provides interest to the architect and engineer.

The recesses in the soffit provide excellent opportunities for using services and special lighting effects. The formers devised by manufacturers can vary from system-oriented glass reinforced plastic and polypropylene formers, modular in size and designed for use with supporting systems, to bespoke units tailored to special situations and functions.

7.3 ADJUSTABLE STEEL PROPS

Steel adjustable props are available from many sources, but care is required to ensure the selection of a satisfactory type for hire or purchase. Diagrams have been produced indicating the range of heights and load capacity of the majority of commercially available props. The practice of using tubular scaffold to provide a supporting system for formwork is only economical where storey heights of over 7 m are to be formed such as those found in warehouse or public buildings, the range of telescopic props available being suited to heights below this, infinitely adjustable through the range of size from 1·75 m and up. Allowance must always be made for any supporting plate at the foot and depth of the lay support members and sheathing at the head.

Some types of props also provide a built-in support for working platforms below prop head level. These supports and the tie frames which are incorporated in many systems ensure that a discipline on plumb and spacing is maintained, both factors of which are critical to prop performance. In reaching a decision as to the type of prop to be employed, it is necessary to visualise the process of form erection for a typical reinforced concrete floor. After the grillage has been prepared and sheeted out, as previously discussed, the site engineer and trades foreman level through using a dumpy level and staff which refer to the given datum level; this calls for adjustment by an operative working below the formwork bay. The adjustment will be to the instruction of the staffman above, and can be best carried out when using props which can easily be adjusted from the slab level or the level of the incorporated working platform. Props which are adjusted by thread arrangements at the head but which do not incorporate a working platform result in time being wasted while operatives clamber to gain access to the adjuster. Threads which adjust at the foot of the prop are liable to become locked by water or slurry which can splash from the previously cast floor slab on to the thread adjustment. The pin which is provided with a telescopic prop should be carefully preserved since a hasty insertion (where the correct pin has been lost) of lengths of mild steel rod can prove to be a constant accident hazard. Apart from the dangers which result from the shearing of inadequate pins a further danger can arise where a rod is inserted causing possible eye injury or laceration due to the projecting ends of some overlength makeshift pin.

When plant is ordered for a concrete flooring contract, care in the selection of the size of props, generally denoted by a number, ensures that as far as possible one set of props will be capable of adjustment to meet the various storey heights encountered throughout the operations involved in a multi-storey building. Small, extra height adjustment beyond that provided by the proprietary prop can be obtained by laying extra plates at the slab level, or by inserting a double bearer system on the underside of the sheathing arrangement. Such arrangements must be skilfully carried out to ensure safety. Methods involving double banking of props to support high level slab arrangements should be avoided because of the considerable side thrusts that result from the intermediate bearer or plate grillages. Problems due to tall storeys or considerable heights from slabs to the underside of decks can be overcome by the adoption of proprietary support systems, or by the use of tube and fittings. Exceptional heights may demand structural steel supports or trestling. Whatever the supporting arrangement adopted care must always be taken to ensure that loads are transmitted axially into standard or adjustable props—eccentric loading causes distortion and puts the plate at head and foot into bending. Forkheads or stripping heads assist in load collection and axial loading and help with the practical process of form erection.

7.4 TELESCOPIC CENTRES

Telescopic centres form an important part of proprietary steel formwork equipment and offer economies on certain types of work, over the more traditional timber joist and bearer systems. Types of telescopic centre most frequently used include those of pressed steel or lattice girder construction and methods that employ a combination of both. The former, while generally of heavier construction, provide an extremely rigid grillage on which can be laid various sheathing materials for solid or hollow floors. Alternatively, centres can be laid close jointed or spaced at tile or hollow former widths so providing a formwork arrangement which can be quickly laid. The centres are usually struck by means of adjustable shoes or nibs formed at the end of the unit which are retracted from the support thus allowing the centre to be freely removed downwards. Wire fingers through the sheathing, and the use of a purpose made carrier, or even temporary scaffold travellers, assist in the striking out of such supports. Where large numbers of heavy centres are used, it may be advisable to use a system of rails or runners which project from the bay such that the centres can be slid out from beneath the newly cast concrete to a position suited to crane lifting. Care must be taken over the regular maintenance of the moving parts to ensure that they work smoothly to facilitate striking. The lattice girder or composite unit works on the same principle, and can also have arrangements which allow a predetermined camber to be set into the centre.Similar arrangements can provide the formation of curved rib bearers to barrel roofs or the haunches of haunch-slab construction. One danger that can arise by the use of infinitely adjustable arrangements in form units is that in which there is a tendency for the settings to vary slightly or be inadvertently altered between uses. Checking against a timber or ply template frame or by measurement from a shaped screed where this is used helps to maintain accuracy. The centres can span either between plates bolted on the side of beams or between previously cast or precast beams, and these are particularly useful for forming suspended slabs over bad ground, or where it is impossible because of obstructions to obtain satisfactory footings for telescopic props. For barrel roof construction, the adjustable centres can be laid to span between the previously cast edge, haunch or crown beam and the area sheathed by a ply sheet or metal ribbing to span between the centres. Care must be taken to prevent buckling of the lattice members, and regular cleaning of and attention to the screw arrangements ensures that there can be great numbers of uses from these extremely flexible formwork components. For the larger spans particular attention must be applied in order to prevent sideways movement of individual lines.

7.5 TIE ARRANGEMENTS

Tie rods or bolts to retain forms in position during concreting were used exclusively for many years. These, quite simply, were passed through the concrete, or over the top of the lift where possible, and while these methods are still effective provided that a suitably substantial bar is used to avoid bending or the threads becoming damaged, present day practice tends towards buried ties and anchors. Precast manufacturers, of course, still employ large numbers of the more traditional tie rods, since tie rods afford an efficient type of fixing in many applications provided that, for general work, a minimum diameter of 12 mm is maintained to reduce the possibilities of deformation during the placing of the concrete or during vibration. For heavy constructional work bright steel ties are used either in simple parallel forms or in a tapered form. These ties can be withdrawn from the concrete to leave a cleanly formed hole which can be filled by 'stopping', followed by a sand/cement mixture punched or packed into place. The better types of tie rods incorporate square or rolled threads which are simply cleaned.

The proprietary-buried tie systems have had a great effect on the processes of forming *in situ* concrete, and the larger precast items, presenting labour savings by spacing forms and avoiding work in straightening and re-threading tie bolts. Welded ties, welded loops and hangers used in conjunction with pressed thread bolts and spacers, whether of timber or plastic, provide an efficient and reliable method of retaining formwork during concreting. The greatest advantage over the more traditional tie methods lies in the positive means of anchorage which is provided for the formwork on succeeding lifts or for the component parts of other form arrangements. Plates for supporting flooring joists and grillage members which retain mechanical plant can be bolted back to previously cast concrete.

Anchorages can also be used to support working platforms and stiffen scaffold arrangements. The cones, which act as spacers to maintain an accurate thickness of concrete when used with ties of the appropriate length, also serve to provide an easily filled hole in the face of the concrete. Ties or loops obviate through-holes in walls and slabs for water retaining or water resistant structures, while ties with weather bars incorporated prevent capillary seepage. The cones prevent unsightly staining of the concrete face by maintaining the appropriate cover and avoiding the rust that can accrue from tie rods in the form of run marks on the face itself. Suspended formwork to structural steel casings can be easily supported by the use of coil hangers, especially where an otherwise expensive arrangement of tie bolts and cleats would be necessary.

7.6 RELEASE AGENTS

Another class of materials which has a considerable effect on the quality of the surface produced by the formwork is that which includes formwork release agents.

TABLE 7.1

Classification of release agents and formwork sealers[a]

Category	Type	Composition	Characteristics	Recommended method of application
1	Neat oils	Without addition of natural or synthetic surface activating agents	May form blowholes. Produces uniform concrete colour	By spray or soft brush
2	Neat oils with surfactant added	Addition of controlled amount of surface activating agent	Low incidence of blowholes. Uniform concrete colour	By spray or soft brush
3	Mould cream emulsions	Emulsions of water in oil where the external phase is oil with a surface activating agent	Low incidence of blowholes. Uniform concrete colour. Reduce efflorescence	By brush or squeegee. Not suitable for spraying
4	Water-soluble emulsions	Emulsions of oil in water where the external phase is water	Low incidence of blowholes. Severe retardation. Dark porous skin. Tendency to dusting	By brush or squeegee
5	Chemical release agents	Chemicals which react with the cement are suspended in a low viscosity distillate	For high quality work and impervious surfaces. Severe retardation can occur with over application	By spray or swab (on small areas). Not suitable for brushing
6	Formwork sealers	Paints, lacquers and other impermeable coatings	These are formwork sealers and not Release Agents. A Release Agent should always be used in the normal way before concreting	As manufacturers' recommendations

| 7 | Other release agents not in Categories 1–5 | Various types including suspensions of wax in volatile solvent | For special applications, e.g. concrete forms. See manufacturers' literature | As manufacturers' recommendations |

a Reproduced from The Concrete Society's Data Sheet, 52.029.

The materials available are classified according to their physical characteristics and chemical composition as shown in Table 7.1.

As with all the previously mentioned materials it is essential that the formwork designer should familiarise himself with these materials. He should be able to give advice on their applications and recognise that certain surface-finish defects are likely to result from an incorrect selection or incorrect application of the techniques available. In many cases it is likely that blemishes attributed to the form design, or the handling and striking methods employed, are really the result of some unsatisfactory practice with regard to the parting agents. Although emulsions and oils are excellent for most commercial purposes there can be little doubt that chemical release agents provide the most consistent results for a surface finish. Chemical release agents also have the advantage in that they can be applied some days before the concreting operations begin and are only activated by contact with the concrete mix. This is particularly advantageous where the time cycle of erection, steel fixing, service installation and such like, are extended due to some complication or revised sequence of construction.

The formwork designer should be aware of the applications and limitations of each type of release agent and, ideally, should note on his formwork drawings the type to be used for a particular situation.

7.7 RETARDERS

The use of retarders is often specified or, if not specified, suggested by the contractor as a means of exposing aggregate. These retarders are surface retarders which are not to be confused with the integral retarders incorporated in the concrete to delay the hardening process as a result of deep lifts or for the prevention of differential deflections. Retarders in paste form for application to the form face have been used for many years and have produced excellent results. Varying degrees of success have resulted on sloping surfaces or where concrete placement has scoured the retarder from the form face.

The recently introduced lacquer type of retarder which can be used in conjunction with a mould sealer gives excellent results. Here the retarded matrix comes easily away from the face with the form and all that is required is a simple brushing out operation. Indeed when lacquer-type retarders are used the concrete hardly touches the form face, and thus the life of the face can be considerably increased.

Chapter 8

FORM AND MOULD MATERIALS 2—TIMBER, TIMBER-DERIVED MATERIALS, CONCRETE AND PLASTICS

8.1 TIMBER

Timber as a formwork material is in almost universal use, and without doubt, every concrete contract utilises some timber form components. Timber is readily available wherever the contract is situated and, labour capable of shaping the material into moulds or forms, can be engaged in virtually every part of the world. Even where steel formwork systems are employed, the designer and manufacturer frequently introduce timber components to form make-ups, stopends and shaped features.

Many large contracts have been carried out where the contractor's method of dealing with the provision of formwork included the placing on site of quantities of suitable timber sections for use as sheathings, props and supporting grillages. This material was cut again and again during the course of the contract until the reductions and wastage called for further purchases, or the contract was completed. The rise in timber prices has dictated the necessity for greater care and control in its use, where previously the material was expendable. Currently, prefabrication is used as the means of bringing about economies in materials costs. Prefabrication also reduces the labour required for fixing and handling. The ease with which timber can be sawn using small hand or power tools when not sufficiently controlled can result in wastage of loose material so that the advantage of prefabrication is that it combats this wastage.

Good quality carcassing timber is the most suitable for formwork purposes. Fifth's quality European redwood or whitewood are the grades most frequently employed. Pacific Coast Douglas Fir provides an excellent material where exceptional wear resistance is required, and other suitable species include Eastern Canadian Spruce. These can all provide predictable service when the loading and modification factors set down in British Standard CP 112 are observed. Designers who are satisfied that the materials can be in use for a short duration may increase the permissible stresses. While it is not necessary to stress-grade materials for use in formwork and mouldwork, timber frequently being understressed in its location within a formwork system, the timber purchased should be free from large or dead knots, waney edges and without shakes. Provided the timber is reasonably air dried, and thus not liable to warp or wind, it should be suitable for sheathing or framing using sound constructional methods. Where edges, arrises and feature fillets on form panels and moulds are to be subjected to a great deal of scour or wear, it is advisable to enter into extra expense and build in hardwood parts. Home-grown Chestnut is an extremely economical material provided that it is straight-grained and sound, and is not likely to stain the concrete. Galvanised screws or nails used for fixing purposes will prevent corrosion which could lead to subsequent failure of the joints. Where particularly fine detail work is to be transferred to the concrete face, this presents another application for hardwood since it will remain stable in small sections. Keruing is another material which is capable of providing maximum re-use for a reasonable initial outlay.

To ensure that the fullest value is being obtained from timber which is used as a form material, it should be framed into panels by nailing or screwing to substantial backing members which are capable of transmitting the stresses that develop within the form to the clamp, tie or supporting system of props. It is wise to adopt common mill sizes for sheathing, joists and bearers and to standardise on these throughout a firm or works. Sizes generally adopted for sheathing are 100 to 200 mm wide boards. Joists are generally of the order of 100 mm × 75 mm while bearers, the main supporting members that collect the loading and transmit it to props and supports, are usually 150 mm × 75 mm and 150 mm × 50 mm. These sizes are the nominal sawn sizes, and allowance should be made for machine waste. The designer frequently selects the size of member for constructional reasons, e.g. the provision of some suitable lap, and many members are used in situations where tradition dictates the method just as much as those of mechanical criteria of design. This is instanced in the situation where 100 mm × 75 mm used as studding, or as a soldier member or joist is layed or fixed with the 75 mm dimension normal to the sheathing to provide a sound ply lap, board joint or bearing. For circular and geometrical work the size of individual boards is governed by the profile, and the specification which governs the size of plane facet allowable on the face of the concrete. Maintaining a common dimension as in the case of the 75 mm dimension which is shared by the joist and bearer materials, can produce economy, since offcuts from form materials such as runners and joists from one part of a contract will thus be able to be re-used in subsequent operations which call for short lengths (i.e. cleats or bearers) to beam soffits and sides.

Imported timber sizes vary and it is advisable to specify sawn sizes from the supplier who then maintains the specified sizes by resawing where necessary prior to delivery. To avoid irregular soffit or wall sheathing, it is essential that all secondary supports such as joists and cleats should receive this attention. Where materials are prepared, they should be brought to a uniform section by 'thicknessing'.

Timber joists of 100 mm × 75 mm section prove most satisfactory when laid with the larger dimensioned face bearing on the primary supports as well as a greater area thus being presented for supporting modular panels or ply sheathing. There will then be less demand for the repositioning of joists, often brought about by creep of the sheathing panels joists as laid. It is possible to avoid the cutting of joists to nett lengths by inserting timber pads or make-ups between bays of proprietary panels or fabricated timber panels, and by moving the joint lines in adjacent panels to allow the joist members to pass. In the same fashion bearers or runners should be allowed to pass and oversail the supporting props upon which they bear. The smallest offcut from timber, i.e. 100 mm × 75 mm, is expensive from labour and material points of view and with the large quantities of such members employed in concrete construction the cutting wastage can rapidly offset savings brought about by improved handling methods, bonus systems and the like.

Timber formwork suitably framed and soundly manufactured can provide many satisfactory uses. Instances have occurred where softwood boarding with glued joints has been used one hundred times and has presented a concrete face finish within the specification of that produced by wrot formwork. Boards glued and thicknessed in panels, screwed to cleats or bearers from the back of the bearer, offer a smooth, clean surface capable of casting a face which can be free from board marks, and which thus reduces the amount of subsequent dressing. Bearers to timber made up into panels should be as wide as the board used in the panel and heavy gauge screws should be used. No. 12 gauge or larger screws ensure that the head is not liable to sink into the bearer in use with the possible consequent loss of fixing. Nails, when used, should pass through both boards and bearers and be clenched at the back of the bearers.

Rips or make-up timbers used to widen a timber face should, where possible, be inserted between full boards in the panel. This ensures that a good fixing is obtained without the tendency for the make-up to twist off when pressure is applied as can occur when column forms are cramped up. Feather-edged timber sections must be avoided as they present a course for infiltrating grout. Where chamfers or bull-nosed corners are required solid features cost little extra to provide. Such features ensure freedom from expensive rubbing down and dressing operations after the form arrangements are struck.

For many years now contractors have been forming striated surfaces by means of timber fillets on the form face. Endless waste of both labour and materials still stems from a poor appreciation of the most suitable means of construction, by avoiding feather edges and by reducing grout infiltration and form deterioration.

In mouldwork and good quality formwork, rebated timber can provide a recess in which such features can be bedded, and thus avoid the grout loss and infiltration which comes from poor joints. Combined with cramping action this arrangement effectively prevents grout leakage.

Timber forms often deteriorate more between uses than when actually used in the casting of concrete. Formwork laid back against irregular scaffolding, or resting on uneven ground, can rack and wind particularly under varying weather and light conditions which can prevail on site. Joints which tend to open under dry conditions form pockets that trap infiltrating grout which prevents the closure of the joint at time of soaking prior to its re-use. When boards absorb moisture from the concrete they tend to expand; due to the hardened fines in the joint, however, they are forced to cup or bow with resultant deformed concrete faces. Straining of the fixings often means that formwork has to be remade.

Timber panels should be stored flat and piled to stay flat during hot, dry weather. Form panels and moulds must be cleaned and oiled immediately after they are struck and before the adherent concrete has hardened which would otherwise mean scraping from the face with resultant scoring. Where these precautions are disregarded, furring of the grain and poor face can result in the succeeding operations.

8.2 PLYWOOD

The great advances brought about in plywood manufacture through its use in aircraft construction and marine work have meant that a plywood now exists which has suitable water resistant properties, and which can be generally made available for concreting. Such plywood can provide over thirty uses per face, and can offer subsequent uses in the casting of ground beams, pile caps and those situations where sawn formwork is specified. Douglas Fir plywood provides a cheap and resilient face material which can be purchased in several grades. Finnish birch-faced plywood is also available which has a slightly more dense face and greater scour resistance. When plywood is used as sheathing, it is advisable to avoid plugged defects which tend to leave marks on the concrete face. With the current adhesives now available, delamination of plywood has been obviated, the glue-line now being completely water resistant and it is likely that the veneers of which the ply is composed will prove more liable to damage through contact with fresh concrete. Damage to the veneers can be combated by treating the edges of ply boards with a suitable preservative such as a lead-based paint prior to the applications of mould oil coating.

Manufacturers produce various solutions for treating ply, and the proprietary treatments are exceptionally efficient in providing the maximum re-use value from the form materials. Orange shellac in methylated spirit, coated twice on most timber or timber-derived sheathing materials hardens the face and reduces grout interference. Other materials for the treatment of formwork, and in particular the edges of ply boards, are cellulose-based primer sealers such as are used on motor-car bodies, or even lead-based primers. These materials can be simply brushed on to the boards in the stack under cover. If damp materials are treated, the coating may fail to bond satisfactorily to the form or mould, and thus become transferred to the concrete face during the concreting operation. Resin coatings and enamels set off by a catalyst provide a satisfactory protection for timber, or timber-derived materials and possess scour-resistant properties which are particularly useful in fighting wear caused through internal vibration or by the striking operations. Plywood for formwork purposes must be selected for its thickness of face veneer and hardness of face. While Douglas Fir plywood is a relatively cheap facing material for form and mouldwork, it may prove economical to invest in a hardwood-faced ply of suitable bonding specification to ensure that a higher re-use value is being obtained. The dense face veneers of Utile, Makore or good quality Gaboon present excellent casting faces. Douglas Fir plywood fibres become scoured from the face to leave a marked grain pattern which transfers to the face of the concrete. This also makes a key for the adherence of grout particles to the form which can result in the tearing of the concrete face. Plastic-faced and impregnated plyboards do of course eliminate these problems, as will a suitable mould coating.

Above all, ply must receive the greatest attention during striking so that edges and corners are not damaged to the extent that will limit the life of the material; plyboards should be cleaned and oiled in the same fashion as timber panels, i.e. by brushing with a stiff brush and only locally scraping with a wide hardwood scraper to prevent lifting of the grain with resultant deterioration of the face. Plywood is generally used on a fixed backing system. This helps to prevent damage to edges and corners occurring, as well as ensuring that the sheet material is not cut into small pieces, thus detracting from its overall re-use value. Trimming of the edges in bulk after a given number of uses greatly assists in maintaining good quality finishes on the concrete face, as does washering of tie bolt holes and careful plugging of unused holes to reduce tearing of both the ply and the concrete face. Internal vibration can be extremely damaging where the ply is scuffed by the vibrator shoe. Small nibs of concrete form in the resultant recess during subsequent uses and tear further parts of the ply face. Such burns should be patched as soon as they become evident by the insertion of plugs and joiners, or by the use of the various chemical fillers now available.

It is the practice of some contractors to use loose sheets of ply for sheathing floor soffits, laying runners and joists as a supporting system and sheathing out with individual sheets of plywood easily handled by one operative, in this case plyboards being lightly nailed to the joists and striking being more easily carried out using this method than with heavier framed panels. Fingers of wire help to preserve the sheathing materials even though the remainder of the system may be removed by crash striking. It is generally advisable to frame up wall and column forms, this being carried out at time of first erection.

Plywood is a suitable material for the production of curved surfaces, since unlike plane surfaces it eliminates the number of joints needed in the form face and facets on the concrete face of barrel vault roofs and similar single curvature work. Plywood can be formed to a fairly tight radius, and where greater curvature is required than the bend which can be induced into one thickness of sheet, two or more thinner sheets can be used, glued or stapled together, to provide the bend required. Good quality curved or geometrical facework can be produced by a suitable thickness of plyboard fixed back to sawn timber or shaped steel ribs.

Mention has been made of the reduction in face joints where ply is used, compared with cases where timber sheeting is employed. Consider a 2400 mm × 1200 mm sheet of plywood and its equivalent where fabricated timber panels are used. The sheet of ply of approximately 3 m^2 in area can be carried and laid by one man, the equivalent timber panel would require two men to handle it. In the timber panel, particularly where unbonded boarding is used, there are some 16 m of joint which are liable to open or close according to variations of temperature and moisture content, and the six or eight boards would be likely to cup or bow with movement of the timber. Grout can infiltrate into such joints and destroy the face quality after very few uses.

In precast and prestressed work, the use of the thinner plies on a board grillage results in a high re-use value being obtained from initial expenditure on materials. For the type of work offering multiple re-use the hardwood faced plies offer better value than the cheaper softwood equivalents. The stability of the sheet material considerably assists in the maintenance of accuracy of dimension of the concrete product.

Impregnated plywood presents the most durable face of all the timber-derived materials although the initial cost of the material will limit its use to the types of work offering great re-use value. The material is extremely hard and frequently calls for the use of tipped tools in its manufacture into form components. Recent innovations in plywood include surfaces into which a veneer of plastic is fused or pressed. The faces thus are completely impervious and in certain instances are advertised as requiring no oil treatment prior to casting. Plywood is also available, impregnated or coated, with a parting agent ready for use in casting concrete. In certain cases the supplier states that these plies need no further treatment prior to re-use. It has been the author's experience however that it is beneficial in most cases to apply a fine film of compatible oil or emulsion prior to re-use thus ensuring that the best re-use is obtained.

8.3 HARDBOARD

Hardboard, another timber-derived material, is an extremely useful material for forming high quality concrete finishes over a limited number of uses. Plain hardboard deteriorates rapidly when in contact with wet concrete, although this can be reduced to some extent by treatment with shellac or a similar water-resisting agent. Oil-tempered hardboard is available which can withstand continued immersion in water having a hardened face resistant to scour and striking damage. If this material is used as a lining material on solid timber backings, the number of uses can be increased and the resultant face finish will be found to be equal to that provided by sheet steel mould faces. Attention to the edges of both types of hardboard is required, and exceptional care is necessary to avoid grout infiltration between the hardboard face and the backing material which can cause damage to the sheathing due to tearing on striking. Hardboard is useful in the formation of curved and flewing surfaces. The cheaper grades of hardboard prove to be economical facing materials where the casting of non-standard sections of an otherwise repetitive concrete contract are concerned. Hardboard can be used on rough backings to stool out moulds or forms as required, the rough form being broken away from the concrete at the time of striking. Sheathing for bull-nosed stair risers or single circular or shaped columns can be economically formed in hardboard. Irregular openings in floor slabs can be formed using hardboard on shaped ribs, the assemblies being broken away on striking. Because of the fine grain-free nature of the material it can be used in the manufacture of stopends on long-line stressing beds, and it can be slotted at close centres without fracture and resultant grout loss. Hardboard stopend liners pierced with accurately located holes can be used to position projecting bars which pass through clearance openings in the stopend framing. Twin stopends spaced by timber and packed with sawdust, sand or polystyrene make ideal separators between adjacent units on the line.

8.4 PARTICLE BOARD

Chipboard and boards manufactured from recycled waste can be used as form or mould sheathing. The durability of the face is a function of the resin or binder content of such boards which should be checked in each instance. The problems are caused by the scouring action of concrete as it is placed and vibrated since these can seriously effect the smoothness of the face and thus the finish achieved on the concrete.

8.5 PLASTER AND CONCRETE

Plaster is an extremely versatile mould material which has been used a great deal in the formation of complicated forms. For single castings of highly featured concrete work the plaster is broken and scraped away from the finished concrete face after curing has been effected. It is from such procedures that the use of concrete as a mould material has been developed. It has been proved that it is capable of casting heavy structural members cheaply and efficiently, as well as providing economic moulds for the casting of detailed precast concrete units. Hardened concrete is a dense inert material and due to its plastic nature can be cast to any profile given a suitable form whether fabricated from timber, cast in concrete or horsed or screeded direct on to a prepared concrete surface. In its simplest form of sleeper walling, concrete can provide a means of ensuring straightness of a precast unit. It can also provide a one-piece indestructible pallet, accurately laid using a level and which will resist the scatter brought about by crane handling of the cast product.

Concrete which is used to form profiled faces provides a relatively cheap solution to the problem of providing large quantities of mouldwork where programmes are tight and many castings per day are required. The construction of concrete moulds is illustrated in this book and it will suffice to say that the concrete constituent of such a mould must be comparable with that of the concrete to be cast within it. The concrete of the mould must be placed, properly vibrated, matured and dressed in the normal way for precast work to obviate suction and adherence between the unit and the mould.

Concrete is readily available on the construction site and in the works and often use can be made of concrete which would otherwise be wasted in casting small moulds.

8.6 PLASTICS

Plastics sheet both rigid and flexible offers the solution to many problems that are encountered in the production of high quality face

FIG. 8.1. Removing the stripping piece from a G.R.P. gulley mould.

Fig. 8.2. The relative lightness of G.R.P. moulds allows one man to strip a unit from freshly cast concrete.

finishes. Some types of rigid sheet can be heat- and vacuum-formed to provide profiles which would otherwise be very expensive to form in timber or steel. Specialists in heat-forming plastics weld up individual profiled panels of plastic facing material to give excellent castings in repetitious work. Formed sheets brought to profile by pressing over a master pattern can be used as a sheathing material when screwed or bonded back to a continuous backing. Large voids between the facing and backing members should be packed with plaster, sand or local timber blocking pieces.

Flexible plastics sheet forms a useful lining material for application to mould faces and forms a glass hard surface finish where self-cleaning concrete finishes are specified. Where such sheets are in use, no open joints or simple lap joints should be left for grout infiltration which could trap the sheathing and tear it from the form face on striking; the material should therefore be lapped over the edges of the form panel, wrapped round battens and nailed or stapled into position. Sheeting which has been used in this fashion has provided interesting finishes to precast concrete when laid over a sand moulding, and thus provides free-form finishes of architectural value.

Glass reinforced plastics, while initially expensive materials to provide, form a material of remarkable lightness, of high impact resistance and are reasonably resistant to scour and corrosion. They can be manufactured into a flexible mould or form with sufficient rigidity to support concrete, yet retain sufficient resilience to allow springing from the cast concrete. As with formed plastics sheeting, it is necessary to provide a master unit or male former around which the laminator forms the mould. Where sufficient lead can be taken or built into the units to be cast, it is possible to incorporate face and returns or pallet, sides and stopends into one mould unit, this being supported on a light framing of steel or timber so designed as to allow advantage to be taken of the flexibility of the laminate in striking.

8.7 ALLOYS AND OTHER MATERIALS

Light alloys of various metals are available from foundries and can be used where particularly complex features are required in all types of concrete work. Patterns can be produced from which the foundry provides castings that can be bored and fettled to the required state for fixing within the timber or steel forms. Complicated anchorage details and recesses can be moulded from castings fixed with the form panels. Repetitive features formed in this way present clean, sharp details on the face of the concrete, wire brushing serving to ensure cleanliness of the finest detail. Re-sale value of the scrap material may assist in the economics of providing such items.

While the majority of materials used in the manufacture of formwork and mouldwork have been covered in these sections, many other materials are pressed into service to meet particular problems. Casting resins and polyurethanes, rubber sheet and formers, building paper, cardboard, metal foil and plate glass can all be used to advantage in providing finishes and features of varying quality. Even linen, perforated zinc sheeting and man-made fibres can be used in the formation of mould panels where vacuum processes are incorporated. Other materials will be mentioned in later chapters where specific requirements of surface demand their adoption.

8.8 FORMERS FOR HOLLOWS AND THROUGH-HOLES

Proprietary arrangements of expanded metal, paper and foil are available to form holding down bolt sockets, and are further components which are available from formwork component specialists. Sheet expanded metal also provides an efficient sheathing material where no-fines concrete is being cast.

An extension of formwork technology lies in the range of ancilliary equipment which is available to those concerned with forming ducts, voids and through-holes. Materials included are steel, foil, timber, plastics, rubber and paper and various combinations of these materials. Foamed or expanded polystyrene and polyurethane form one branch of materials, although frequently as only one use is taken from these materials their apparent cheapness may prove to be misleading. Concrete can be used to form virtually indestructible formers, and can be simply cast using techniques with which site personnel are familiar and thus offer considerable economy.

Steel formers for hollow concrete units are of several types. One particularly efficient arrangement, commercially available, is that in which preformed sheet steel or alloy is retained in position during casting by extruded plastics stripping fillet, the removal of which allows the former to spring from the concrete face and present clearance to allow removal from within the unit.

Fig. 8.3. Inflatable formers can be obtained in a range of sizes from prestressing duct sizes up to culvert sizes.

Fig. 8.4. Large pneumatic void formers were used to form these culverts.

Patent techniques of hollow formation are legion; two techniques are particularly worthy of note, in each case the patents are the property of British companies. One system utilises resins mixed within a plastics sheath while the plastics is passed through a conforming chamber, the result of the process being a semi-rigid former of disposable nature for casting into precast flooring and similar units.

Another system depends on the introduction of stressed wires which are in turn wrapped with wire mesh and plastics sheet. In this instance the system may be used as a 'lost former' method, or the formers recovered at will.

Solid sections of expanded polystyrene or lightly framed timber assemblies may be used, although in all cases where formers are submerged care must be exercised to combat flotation.

This tendency to float is particularly evident where inflatable cores are used although simple foil tubes and formers have been displaced by flotation in the manufacture of prestressed components. It is essential that care is taken to restrain the formers and to provide tell-tales to enable checks to be made on duct positions during the concreting process.

Chapter 9

THE MANUFACTURE OF FORMWORK AND MOULDWORK

In this chapter it is proposed to outline the arrangements for the large-scale manufacture of formwork and mouldwork. The arrangements are those employed by the specialist manufacturer who has a shop devoted entirely to this type of work. The materials discussed are timber and steel, since the majority of purpose-made forms produced by building, civil engineering and precast manufacturers in their own works or on site use these materials with additional proprietary or purpose-made fittings, fastenings and ties. The provision of forms and moulds constructed entirely of steel is the province of a specialist engineering firm, the work involved generally being sublet by contractors to such a firm. The methods outlined here are elaborate but are reproduced in a simpler form in all manufacturing, whether in temporary carpenter's shop on site, the contractor's plant yard or the mouldshop of a precast concrete establishment. The machinery described is necessary where good quality moulds are to be made on a production basis with hand assembly work cut to a minimum. Simpler arrangements may be established with perhaps a bench saw and a few powered hand tools, burning and welding gear, since simple arrangements like these can produce efficient and accurate form and mouldwork. The size of the shop is tied to the decisions reached at the planning stage with regard to place of manufacture of the formwork where *in situ* work is concerned, and the volume of new mouldwork required over a given period in the case of precasting.

9.1 TIMBER MOULD AND FORM MANUFACTURE

The arrangements discussed in this chapter have been based on the procedure adopted in joinery manufacture to which prefabricated timber form and mould construction is closely allied. The personnel required in this procedure have their counterpart in the joinery works. Form and mouldwork is essentially the province of the tradesman who fits his basic material to the formation of concrete, and the requirements of the finished concrete which controls the methods used in the construction of the timber or composite mould. Beyond all else the critical requirement is the consideration of the re-use value consistent with the appropriate standards of accuracy and finish with the minimum of refurbishment.

The departmental manager deals with the forward programming, production planning and possibly the purchase of material, and is primarily responsible for liaison with the customer or parent company. Details of the order, delivery dates and specification are conveyed to him by the works manager. The foreman is responsible for the day to day running of the mill and assembly shop while his respective charge hands supervise the labour employed. He is generally responsible for engaging the labour force in the shop and is required to ensure that they are suitably instructed in the methods and practices generally adopted in the works.

The foreman is assisted by tradesmen in matters relating to the accuracy of setting out full-size details and 'rods', and the marking up of timber in accordance with the detailed mould or formwork drawings. The works requires certain basic machinery, such as that employed for sawing and planing, together with additional plant dictated by the nature of the work.

Cross-cutting work calls for a substantial machine, preferably one capable of accommodating trenching heads, and it is useful, though not essential, to have a canting head machine for bevel and splay cutting. Small but efficient canting head cross-cut saws have recently become available—particularly from Europe. The trenching attachment is particularly useful when forming housing joints at stopend positions, and when used for cutting splayed ends. It is also useful in producing boards which are required to be split to allow striking from between returns or features, where splay cuts at the ends of I-section beam features, and transverse duct positions are to be made in quantity and where anchor recess formers are required in bulk. The sawbench, of suitable size, may either be manually or power fed, though obviously where production is con-concerned the power-fed machine ensures constant turnover of material throughout the shift.

A planer and thicknesser may be combined, but with the long lengths now required in most form and mouldwork, it is advisable to employ separate machines. The surface planer will be chosen for length of table to facilitate straightening of long lengths, while the thicknesser preferably should have a sectional roller feed.

A substantial drill press capable of driving large diameter bits is required for accurate boring of duct holes in formers and dowel bar holes in stopends. These machines, together with a bandsaw which has roller guides, can complete the basic machinery lists when supplemented by powered-hand tools such as drills, saws and hand routers. Other machines may be required when any departure is made from the plane forms and simple box-type moulds, or where quantity production is demanded. A versatile machinist can produce from these various machines a great number of timber sections with a variety of bevelled and splayed cuts. Grooves can be produced by means of a wobble saw. By working on the edge of the planer table rebates can be run and by utilising the tilting fence of the planer, various feature fillets and mouldings can be produced. The bandsaw is used to prepare shaped blocks for striking fillets, stopend checkouts, and shaped ribs and formers for simple curved work which are cleaned up when required by hand on the bench. At a later stage a dimension saw will be needed to accurately cut both solid material and plywood sheet. A spindle moulder with suitable blocks and cutters will be necessary to run the more complicated features and to combine bevelling with rounding and rebating, for instance where features cloak the edges of plyboard in raised facings. The spindle moulder with a collar may be used to finish machine-shaped ribs on curved work, while a router is required to form inside splays or openings and housings for features inset into the form face.

Another machine not frequently met with in mould or formwork shops is the woodworking lathe. It is however an extremely useful machine where tapered pins and plugs are required for through ducts in prestressed concrete work. Turned pins are also useful for forming holes in diaphragms cast to position stressing-cable assemblies during casting. Pellets for insertion, where screwing through the face is employed in sheathing fixing, can rapidly be produced in stick form from offcut material. The hollow chisel mortiser proves to be an asset where framed assemblies are required as in the use of backing arrangements to prefabricated forms. With modern adhesives and a well-made housed joint the need does not often arise for the adoption of classic joints which are used in general joinery work. The manual or air-operated machine capable of driving a variety of staples and nails in one continuous operation is extremely useful. Staplers are perfectly suited for the fixing of ply facing to sheathing, or ply sheathing to timber-backing members, the consistent depth to which the fixing is driven helps in rapid filling with a plastic stopper prior to sanding operations. A hand-sander, of the belt or pad variety, speeds up the final cleaning of mould faces or the rubbing down between applications of surface hardeners or finishes. Portable sawing and planing machines are used more in refurbishing work than in the original construction, the saw being a particularly useful piece of equipment for reducing base or side panel widths as modifications are made.

It is essential that construction be carried out in such a way that increased productivity results from the application of planned methods of assembly.

Accurately machined components which pass in a steady flow from the mill to the assembly shop with only the fitting and fastening operations to be carried out on the bench to produce the finished mould or form, quickly repay the initial expense of the installation of machinery. In this way the carpenter's and mouldmaker's efforts can be applied to produce the extra finish and soundness of construction required in moulds and form panels necessary to meet the rigours of constant re-use without refurbishing.

The machine shop which is attached to a precast works will provide a service in the supply of shaped fillet spacers, wedges and stopend blocks which can, in the main, be produced from offcuts of material prepared for mould construction. This service may come under the control of the mill foreman or chargehand machinist.

The mill foreman primarily will be responsible for the preparation of material to the stage where the mould construction becomes a matter of pure assembly work, carried out by screwing, nailing or bolting by mouldmakers in the fabrication shop. He is responsible for the labour that handles timber into the shop and the machinists who work to the cutting lists prepared by the setter out. He is also responsible for the progress of the timber through the shop, its cross-cutting and sawing, and in the preparation by surfacer and thicknesser, bandsaw spindle moulder and router operations through to the fabrication shop, and finally, carpenter's shop or mouldshop. The marking out of the timber is carried out by the 'setter out' working from full-size 'rod' and mould drawings, and he is responsible for ensuring that every item of timber is duly marked out in accordance with the cutting list and details.

The setter out will examine mould details when received from the design department and produce such full-size details as are required to indicate sections of timber, special end treatments and feature formations, and then produce a cutting list for each mould or group of form panels. These lists will be used throughout the mill and shop as instructions on numbers, lengths, sections and the preparation required. Copies will also be used for stores control and costing purposes. The setter out should also list all supplementary parts and requisition these from the stores. The primary function of the setter out is that of maintaining the accuracy of the product and as such is responsible to the shop foreman. He raises questions of detail with the person responsible for the design, and this close contact with materials control, and detailed examination of all drawings while preparing rods and patterns, provides the equivalent of an inspection service within the shop and mill.

The mouldshop or carpenter chargehand is responsible for checking materials as received from the mill. He also draws up from the stores the necessary fastenings and fixings for assembly purposes. He works on prototype panels or assemblies with the carpenters or mouldmakers and checks the work as it proceeds, liaising with the setter out over the various constructional details. Where several pairs of carpenters are engaged on individual parts of a mould, the chargehand is mainly concerned with the fit of the parts and the incorporation of the parts into a finished mould.

The chargehand carpenter sets up jigs and templates for runs of similar panels and also supervises the carpenter labour engaged in the refurbishment or modification that is carried out on the previously manufactured work. When all the assembly work has been completed the shop foreman and setter out will give a final check on the mould for dimensional accuracy and soundness of construction, and ensure that the positions of dowel holes, inserts, features and pads are as required for the first casting operation. Where the manufactured forms are despatched for use on site, this may be carried out during the first set up on site. With geometrical or complicated work it is advisable that the trades foreman from the site, and key operatives concerned with the use of the components, should be able to see the arrangement assembled in the shop. At this stage the order of assembly and methods of application may be discussed. With large-scale work it is advantageous to have the chargehand from the site work on the trial erection in the shop so that he can familiarise himself with the labours involved, and be able to instruct his own gang during the first set up on the site.

In precast works moulds are handed over to the inspection department for final clearance for casting. This inspection which is carried out using the concrete unit drawings serves as a check on all mould-manufacturing processes from the preparation of initial profile through to the assembly and preliminary set up.

The foregoing has revolved round the set up maintained by the specialist mould manufacturer and the larger precast and prestressed concrete establishments. There are also a great many smaller shops of a temporary nature which are set up on site by contractors engaged on concreting work. These are usually controlled by the trade foreman and utilise the minimum amount of plant in the form of a sawbench with perhaps a few powered-hand tools. With these limited facilities a great deal of useful prefabrication work can be carried out. Boxes for kickers and columns can be prepared, beam sides and form panels machined as well as stair strings, riser boards and similar items. Care in the purchasing of suitably machined board and such items as machined sections of fillet to be used in features can make for ease of manufacture. In this way the more complicated work in producing chamfered arrises and similar work can be carried out by the material supplier. Wherever possible, full-size setouts should be made on sheets of plywood which are intended eventually for sheathing as these will serve to maintain the accuracy of site produced formwork.

Where the size of the job merits expenditure, it is useful in multi-storey work to have a powered-bench saw at the working level ready for use in providing the spacers, rippings, blocks and wedges required in the erection of floor and wall formwork. In this way a contractor can ensure that full use is being made of the timber on site while offcuts from larger timber joists and bearers can provide the material for the accessories. A mobile saw can also be used for reducing the depth and width of column and beam sides and for the trimming of ply sheet edges which have become ragged and damaged through constant use. Stapling machines are useful pieces of site equipment especially where plyboards are used for sheathing purposes. Wherever machines are operated on site, particular care must be applied in matters which relate to safety, fencing, power supply and such like.

The mouldshops of many precast companies are similarly equipped to a site carpenter's shop, with perhaps the addition of a combined planer and thicknesser, and the versatile bandsaw. The mouldmaker in such a shop prepares his cutting list in co-operation with the chargehand or foreman and then machines and prepares the material as he requires to bring forward the work on the bench. Obviously, by working this way, mould costs are higher than those where production methods are used, but the individual attention given to the mould product is often repayed by the multiple usages obtained on the casting bed. Where extremely accurate work is concerned zinc templates or ply templates are prepared for both the unit and the mould. These templates are used in checking moulds and units during the course of the contract.

Where form and mouldwork is required on site or in the works of many contractors, the services of the contractor's own joinery works are enlisted. The greatest problem that arises from this is that of using bench joiners to produce formwork, and the question of whether it is possible to obtain work of a quality appropriate to the use to which it is to be placed. Operatives who are used to producing work of joinery quality require careful instruction to ensure that they manufacture in such a way as to produce moulds and formwork at an economical rate. This particular problem does not arise with machinists' work where the standard is automatically set by the number of finishing operations carried out on the material. The hand workers however need to be instructed that they should omit the final finishing operations which are usually involved in joinery quality work, while ensuring that the production of formwork and mouldwork is soundly manufactured for multiple re-use without the need for over-fettled surfaces. The problem here is that of instructing skilled men to produce components of a lesser quality than that which they would normally maintain.

9.2 STEEL MOULD AND FORM FABRICATION

It is now necessary to consider other materials currently used in construction. In a changing economic situation with varying costs of materials, the emphasis in materials selection changes radically over a relatively short period. It is rapidly becoming an economically viable proposition to use steel in relatively simple applications and where, perhaps, fewer re-uses are available than would have been required to make the selection of steel profitable only a few months ago. Steel forms have for some time now been the province of the civil engineer and precaster with either complicated sections or applications offering considerable re-use. Today simple steel forms can provide satisfactory returns not only on relatively small numbers of uses, but in such simple operations as those connected with column and frame component production in building.

It is not intended here to discuss the establishment of a works facility such as those usually employed by the main steel form and mould manufacturers. Indeed very few contractors or precast suppliers could afford to equip themselves on the scale of those that are foremost in the field of form fabrication and supply of moulds. However some suggestions are laid down for equipment to supplement the normal mould fitter's workshop to enable the manufacture of simple special items, one-off moulds, formwork components and accessories.

In the case of steel forms and moulds the costs of manufacture and the cost of equipping the workshop facility can be maintained at an optimum economic level by skilful materials purchasing. All material items that call for the use of sophisticated or heavy equipment such as shaping, milling and the larger rolling operations should be carried out by suppliers, or sublet to engineering firms who possess suitable equipment. This avoids heavy investment which would otherwise be required in the purchase, housing and maintenance of the equipment.

The fitter's shop generally has a suitable supply of the normal tools and equipment needed for welding, burning, sawing and the general preparation of stock such as plate, channel, box section and similar materials, while most works incorporate a suitable covered area in which these materials can be stocked. A most useful supplement to the burning equipment is a profile machine which can copy templates and cut steel profiles from plates. The simple straight line burner alone can save considerable time in plate preparation and the profile cutter under the control of a semi-skilled man can be very useful in the production of stopends, gusset plates, stiffeners and such like.

With regard to the cutting to length, box sections and similar materials for bearers, soldiers, walings and other members, the continuous power saw will prove to be the most economic piece of equipment since it can produce cut lengths which are free from the burrs and defects usually associated with abrasive cutters and similar tools. Equipment using abrasive discs has however a place in the preparation of the smaller sections of steel and certainly where outputs in cut lengths are high.

Pillar drills, and where possible radial drills, are essential plant and equipment items. Another machine which is valuable in the preparation of the simpler mechanisms for closing forms and actuating equipment is the centre lathe, preferably with a gap bed to allow the accommodation of large eccentrics, bosses, etc.

As stated earlier the larger items of plate involving rolling and machining should be purchased from specialist suppliers, as should any items needed for milling and boring.

Apart from the machinery and tools mentioned there is a further requirement with regard to equipment. The fabrication shop must be equipped with a suitable lifting gantry and plant which is capable of handling both bulk material deliveries and fabricated mould sections. The mould sections may well present problems of scale and shape which outweigh the normal considerations of the mass to be handled.

Whatever equipment is used it is essential that a substantial fabrication bed should be incorporated within the structure of the assembly shop. Welding and fabrication induces massive stresses into the carcasses of form and mould panels. It is essential that there should be sufficient hold down and anchorage points incorporated to enable the assembly fitters to combat the stresses and produce component parts free from warp and wind.

A not inconsiderable part of mould and form manufacture is the operation of producing straight plane forms. This process involves the straightening of members and trueing of plates for which purpose strongbacks and hold downs are invaluable.

As a general principle, whenever a workshop is being prepared either for the purpose of steel fabrication, timber mould manufacture or general concrete production, the introduction of a grillage of sockets or threaded inserts into a sound structural floor slab assists in the many operations of fabrication, machine installation, jig and template erection, trial assembly and similar work.

9.3 PLASTICS WORKING

Where plastic mould fabrication is to be carried out, workshop equipment and tool requirements are far more modest and generally consist of a well lighted open-span structure with adequate temperature control and ventilating systems. Storage of resins and sheet material should receive particular attention with regard to the requirements of statutes, and a study should be made of the Factory Inspectorate's recommendations on these matters.

In all the situations described in this chapter care is required in planning and method engineering to avoid accident hazards. Any process which could jeopardise the health of those involved must receive special attention. Protective clothing, masks, goggles and the statutory supplements such as milk, limited working hours and specified rest periods must be observed at all times.

Chapter 10

CONSTRUCTION OF *IN SITU* WORK
1—FOUNDATIONS AND WALL FORMATION

Setting aside the question as to where mould and formwork construction should be carried out this chapter discusses the various casting operations of a typical *in situ* structure. While obviously the operations discussed occur within general concrete construction, they will be encountered in varying quantities according to the nature of the structure that is being cast. Concrete requires placing and vibration must be applied correctly to ensure full compaction and the exclusion of entrapped air and water which can otherwise cause faults in the final concrete structure and face finish. This vibration demands that particular attention be applied to all joints, bearers and struts in order to avoid grout leakage and settlement of the supports as it is applied. The textbook warnings about nailed wedges and shakeproof fittings in moulds and forms apply, particularly where high frequency vibrators are being used. Flotation and uplift must be considered in all operations particularly where highly workable concrete is used.

10.1 FOUNDATIONS AND GROUND BEAMS

Concrete work begins with the foundation work and comes within one or other of the following categories—strip foundations, cappings to pre-driven pile clusters or ground beams, and part of the oversite slab, or raft foundations. Provided that the excavation work has been carried out accurately to levels and trenches have been well cut for ground beams, the work of forming the ground beams will be a simple matter. Subject to the engineer's approval, material recovered from demolition elsewhere can be used for forming the concrete; even panels that are too badly worn for use on other exposed parts of the concrete structure may be utilised for this purpose. Forms for the work to be carried out on ground beams, strip foundations and column footings can be strutted from the face of the excavation, wired through the concrete or supported by paging from pegs driven into suitable ground. The latter method should be confined to situations where the ground is particularly reliable and is not likely to deteriorate during bad weather. All struts which support formwork must bear on plates or lengths of timber, so placed as to spread the loading from the strut over the largest possible area of earth, to eliminate settlement or movement of the form as a whole. The basic principles of form manufacture should be applied in spite of the fact that the finished face will never be exposed. The use of soldiers and walings behind the sheathing material will result in a smaller number of supports being required with fewer pegs having to be driven, if this is the method of support adopted. Where a crane is available then the foundation forms can be fabricated into larger panels or made more substantial thus reducing the ties and supports required.

As the blinding operations are being carried out, the experienced formwork tradesman should take advantage of the opportunities presented, to provide rigid fixings for form panel toes by inserting small concrete blocks or spacers on the line edge of the slab or footings. After the blinding has hardened, those blocks will provide a purchase against which panels may be chocked by wedging and strutting, and will speed up the whole formwork operation. Blocks can be inserted in strategic positions when casting raft slabs such as to make allowance for the strutting of concrete sections of widths that can be cast without the use of tie rods or bolts. Where it is proposed to strut from blocks which are set in the blinding it is advisable to lay the blocks in lines and span them with a substantial plate. This plate can be suitably wedged and fixed to ensure that the stresses are uniformly distributed over the series of blocks. Masonry nails can be driven to support plates where it has not been possible to drop in blocks or metal pegs, shot fired fixings being quite adequate in most situations. Grooves roughly chopped in the concrete provide poor seatings for props as the stresses will not be equally distributed over the base plate of the prop. Beam clamps can be dropped over the outside of the forms and combined with concrete spacers, while cast-in blocks provide a substantial tie arrangement for ground beams.

Where depth of beams are in the region of a metre or more it is

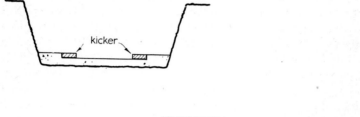

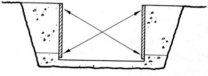

forms to lean fill erected and braced

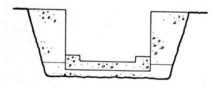

structural slab cast

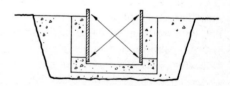

forms re-used to form structural walls

FIG. 10.1. Principles of back casting in ground. *Note:* similar principle can be used to form large section ground beams, pile caps, etc. Steel omitted for clarity.

economical to use a buried tie system. The short bolts used with the system may be easily removed from between the form carcass and the dig on striking. It is advisable when the ground beams are being cast to decide which forms may be profitably removed and which may be considered permanent, the latter being formed from recovered ply sheathing material or corrugated sheeting. Frequently with floating rafts, as demanded in subsidence areas, below good platforms and similar structures, the size of the form component must be considered to ensure that it can be removed from the cast concrete without difficulty. In this instance it may be necessary to shorten the walings and bearers to allow removal through the openings in the concrete which are to be subsequently filled solid.

Back casting is a technique which is particularly helpful in bad ground and where backfill around foundations and rafts is to be carried out using lean concrete. For the back casting operation, the blinding coat incorporating the support blocks previously mentioned is laid and, if necessary, forms are nested on this blinding and the backfilling work is carried out using the forms to generate the outer profile of the eventual structural concrete component. The resultant concrete-lined 'form' is now available for steelfixing and other operations such as tanking or lining work. The area is now clean and can be kept free from water and falls of spoil, so that operatives have the best opportunity of working in reasonable conditions, e.g. setting up accurate steel reinforcement and providing kickers for succeeding lifts and bays. Back casting simplifies form support problems, since it becomes a simple matter to prop between forms rather than attempting to contain concrete with forms, often spaced far apart, using ties or props from the excavation.

When foundation rafts of any great depth are to be cast, a specified time is often required between the casting of adjacent sections of the work. This stipulation can often be put to good account in that the steelfixing, formwork and concreting operations can be phased to afford a continuous flow of work. When compared with continuous casting, a greater volume of concrete can be cast from a given quantity of formwork over a certain period. Skilful use of weekend working as curing periods and the careful timing of the concreting operations can turn an otherwise unproductive period to good account.

Although there may well be a certain amount of extra fitting of formwork around the upper and lower reinforcing steel, this can be localised within the height. This method, worked in chequer-board fashion over a slab, reduces the formwork operations whenever three or four sides of a bay are formed by the previously concreted sections of a slab. Simple edging formwork is only required where three sides are formed by such concreted sections adjacent to the perimeter of the slab. Day joints and construction joints are specified for this type of work and joggle joints are often required as are plastics or copper waterbars. In such cases the formwork must be arranged with horizontal joints in order to accommodate the top and bottom steel in the component and where necessary a groove for the weatherbar. The forms must also be made in lengths which are such that the form can be turned and withdrawn from within the reinforcing steel after the casting has been completed without the need to exert any force on the steel and freshly cast concrete.

During the casting of foundation rafts, particularly those which act as water retaining structures, it is necessary to form lead-ups or kickers for waling. These items spring from the finished concrete level at the perimeter of the slab and are monolithic with the section of the foundation being cast. The soldier members that support the slab edge sheathing should be allowed to project enough to allow the spacing and bolting of a suitably shaped outer form. The inner kicker form is erected when the casting of the foundation operation proper has reached such a level that some bearing can be obtained from the concrete. In this way support for the lower edge of the kicker panel can be obtained when sufficient hardening has been achieved so that surging beneath the bottom edge of the kicker form does not occur. When kickers, monolithic with slab sections, are cast the utmost care must be taken to maintain accuracy with respect to the profile of the succeeding waling surfaces, since on this depends the line obtained.

10.2 KICKERS

Kickers for internal and external walling and columns can be cast as a subsidiary operation, the outlines being formed in some material of suitable depth, spaced from the projecting steelwork of the column and walls. This is best carried out as soon as the concrete slab can take the weight of the operatives and engineers involved. The line and level engineer will drop lines from the profiles on to dots of mortar adjacent to the column or wall, to provide the datum lines from which carpenters can work when fixing the kickers. The height of the kickers should be so arranged as to support the butts of vertical steel while it is being fixed. The kicker should be not less than 50 mm high to allow the succeeding form panel or column box to lap a minimum of 40 mm. This provides a good sound joint which excludes the possibility of grout leakage and honeycombing at the joint.

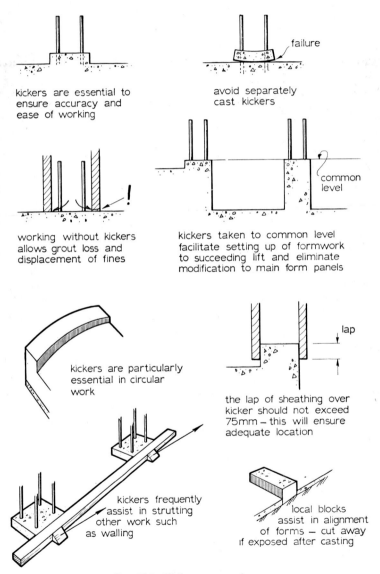

FIG. 10.2. Kicker construction.

Where it is proposed to tie back the foot of the forms by through tie or buried anchorage, the kicker height requires to be increased to about 150 to 200 mm to obtain sufficient cover over the buried anchorage or bearing around the through tie. For circular columns the kicker is best kept to a depth which allows the kicker timbers to be bandsawn from solid material. Higher sections of kicker would, in this instance, involve construction similar to that used for the circular column form and thus prove to be expensive in manufacture. Kicker formation in connection with circular walling requires special attention. Care taken in the provision of sections of shaped formers for such kickers are amply repaid in the initial walling operation. Circular panels which are otherwise awkward to handle, can be readily brought to bear against the kickers, and can be rigidly supported by means of pressure applied at the foot. This support can be achieved either by wedging from buried blocks or props which are punched from the opposite wall face, or by the use of ties or buried anchors. The inner kickers to small shafts, flues or ventilators that spring from the raft should be treated in such a way that they can be struck from within the formed upstand without spalling of the concrete face. This involves the use of keys or stripping fillets to allow the extraction to take place.

When kickers are cast on sloping or irregular slabs the opportunity should be taken to produce a constant level at the top of the kicker, thus providing a level line at the bottom of the next form. With stepped slabs and slabs which have irregular edges this method avoids the need to cut the panel sheathing over steps at the time walling is erected. Adjustments and steps are thus formed while small, easily handled sections of material are being used, and not while bulky wall panels, which have to provide multiple re-use, are being positioned. Where walling springs from ramping slabs, the kickers should be arranged in such a fashion that they provide regular increments in the wall casting height. These increments can be obtained by deepening the form with back boards or sections of sheathing fixed in panels so that soldiers project at the head of the lift. Where baluster walls are being cast to stairways, the kicker for the baluster should be in the form of a cut string which is arranged to provide a projection beyond the nosing line on which succeeding forms clip. Kickers and lead-ups are vital items in concreting work since, when accurately laid and squarely cast, they provide the foundation for the erection of all formwork which springs from the slab or beam level. In circular work sound kickers help to maintain both plan form and plumb.

10.3 COLUMN BASES

Column bases are treated in the same way as ground beams, with regard to the materials being utilised, although they are frequently located deeper in the ground than the beams. Permanent brick or block formwork is often laid, backfilling proceeding prior to, or whilst, concrete is being cast. This method of permanent brick formwork proves useful on asphalt-tanked structures where the coatings of asphalt are applied to the brickwork. Steel reinforcement is placed and the formwork and concreting operations proceed as secondary operations.

10.4 DUCTS AND POCKETS

In groundwork runs of concrete ducting often have to be cast and it is worthwhile providing an accurate form in sections of length to suit the slab castings. This type of form can be dropped into position on concrete spacer blocks and centred-off from adjacent timbering to ensure accuracy of position. Small jacking screws can be used both to strut the duct sides into position and to strike the forms from the position after concreting. The formation of shallow ducts can be carried out as the concreting proceeds since there is insufficient head of concrete to cause surging at the base of the side form of the duct. Where the duct depth exceeds 450 mm, however, it is preferable to form the duct bottom with kickers as a preliminary operation, this being followed up by the concreting of the duct sides prior to placing concrete in the slab.

Where pockets need to be left in slabs or column bases a number of methods can be employed. Timber blocks splayed on all four faces can be wrapped with thin waxed paper. Larger blocks can be fabricated from ply or hardboard with the sides arranged to pass; this enables the separate pieces to be folded back into the space formed and thus can be extracted from the formwork. Concrete formers provide an extremely economic means of forming openings and through, tapered holes which allow withdrawal and which are wrapped with paper or expanded polystyrene sheet so that they can be simply withdrawn by crane. Where pockets are required for holding down bolts, turned timber pegs or bent up boxes of thin

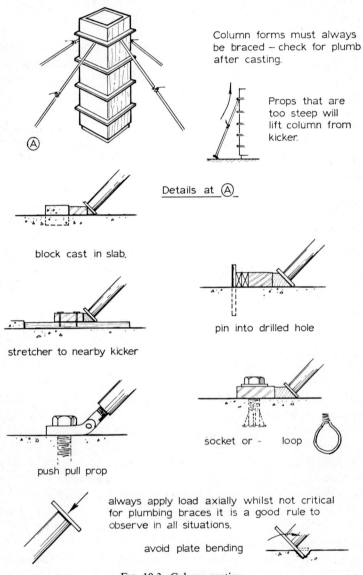

FIG. 10.3. Column casting.

sheet steel attached to timber stretchers provide simply struck arrangements. Proprietary expanded metal cages are used to form recesses and holes and have the advantage of leaving a keyed surface for subsequent grouting. Bolts hung from stretchers require to have a smear of grease on the thread to prevent rusting, and the adhesion of grout to the threads.

Formwork to large sumps and recesses in heavy foundation slabs need as much attention as any other part of the concrete structure. Frequently, hastily formed boxings cost far more to strut and strike than would a properly framed arrangement which on striking yields materials for further use. Where struts pass from back to back of opposing waling pieces, provision of small nailed cleats prevents movement of the strut while concreting is in progress, and assists with the placement of the wedges. Wedges cut from homegrown hardwood cost little more than if they were cut from imported softwood and will last longer and provide for better bearings for puncheons and struts than those in softwood. Wedges must always be spiked to prevent movement during concreting, particularly places where heavy vibration occurs. The small packers and odd rippings often used under heads of bolts or ties are constant sources of danger which include crushing, splitting and falling, since they cause many unnecessary concrete hacking and remedial operations.

10.5 FIXINGS AND FASTENINGS

At this stage in the discussion on site work, it is opportune to consider the fixings that are used by those concerned with the erection of formwork. The nail, which should be a good stout gauge of the chequered, head-wire type, is always used in shear and driven to within 4 to 6 mm of the head in framing work so that it can be drawn by nail bar or claw hammer. Where sheathing is being fixed by nailing, the nails should be driven to provide a dovetailing action with nails alternately angled slightly to the face of the boards. Failure to do this results in a board fixing which easily parts as the panels are struck from the concrete face. Clenching of nails behind bearers is the most satisfactory way of obtaining full strength from a nailed sheathing face attachment, and it will be found that ringed shank nails, cement coated, are excellent for the purpose. A useful device on site is the patent nail drawer which has a built-in hammer action for driving the pincer teeth into the timber, and a fulcrum action which tightens the pincer grip as leverage is applied. For the fixing of sheathing many versions of the stapler provide rapid fixing.

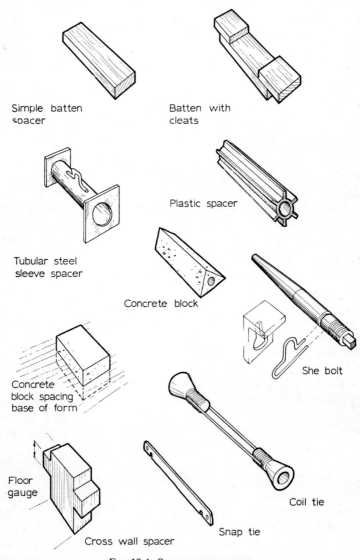

FIG. 10.4. Spacer arrangements.

Screw fixings should always be made using heavy gauge screws of size 12 or 14.

Where large timber sections are being fastened together, coach screws with plate washers at the head provide an efficient fixing, though pre-boring is required to avoid the splitting of the timbers that are being fastened. Framing anchors provide an efficient means of joining timber backing members; when fixed by nailing with clout nails, they eliminate any need for pre-boring and can be treated as a recoverable item when the framing is dismantled. Used singly, they provide a satisfactory corner joint for framed panels and when singly or in pairs provide an excellent anchorage for waling attachment to soldiers. These fixings, together with ordinary black coach bolts, provide the necessary range of fixing for timber forms, although such items as corrugated timber fasteners, toothed plate timber connectors and shear plates, while expensive to provide in the first instance, can reduce costs by increasing the efficiency of bolted joints. These items reduce wear on demountable timber junctions such as folding struts, and the junctions between panels which are held by bolting. Screwed connections should not be dismissed. While initially expensive to make, screwed joints when used in conjunction with glue, and in particular resin adhesives, withstand many uses and thus can prove economical in the long term. Recently, resin producers have brought fillers and pastes on to the market which can be used with confidence for covering nail and screw heads in sheathing work. The compounds mostly comprise a resin and a hardener which when mixed and placed by knife set rapidly to produce a durable plug which can be machined or planed smooth.

For larger holes and defects in formwork which is used on concrete for a commercial surface finish, toothed plates can be driven to fill the major damage or defect, the final filling being carried out by the application of resin.

10.6 WALL FORMS

After the kickers and lead-ups have been formed the next operation involves that of erecting the formwork for walls and columns. Walling can be divided into several categories: that which presents two plane concrete surfaces, that against piles or brickwork required on one side only, and circular or geometric walling. Walling operations may be complicated by the formation of piers that project from the wall, by beams that project from the face and by features, either as projections or recesses, in the concrete face.

As previously discussed, with the height of lift decided, the methods

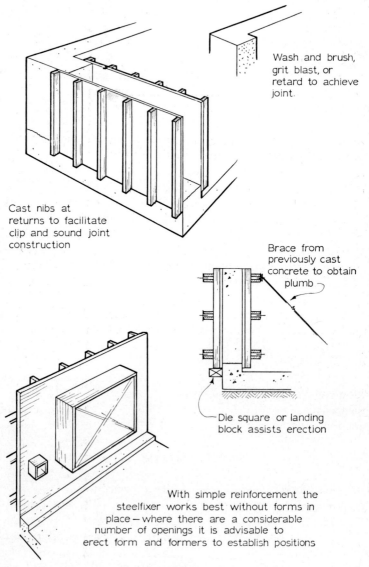

FIG. 10.5. Wall formwork.

of constructing the formwork and retaining it in position during casting are diverse, although the methods of framing must necessarily be related to the tie system adopted. The traditional method of using

FIG. 10.6. These form panels are being positioned on a bridge site. The cable winch provides fine adjustment prior to the final securing of the bracing members.

through-tie rods to retain the form is still widely used, but care is needed to ensure that the minimum number of through-ties are used that are consistent with the formwork strength. Where the full height of the walling is being cast in one operation, and forms are not used

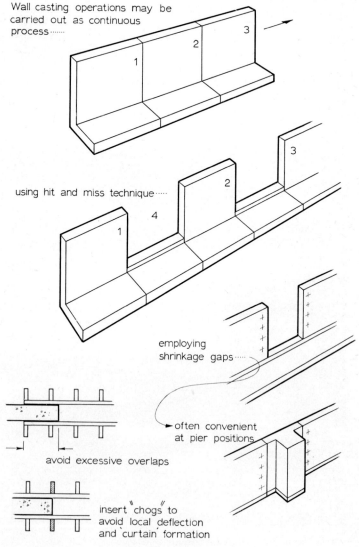

Fig. 10.7. Wall casting.

to cast lift upon lift, the spacing of the tie rods or bolts can be related to the strength of the materials used. Care must be taken to pass the ties through the sheathing which is immediately adjacent to the soldiers, or bearers, to reduce the possibilities of waling deflection which allows spreading of the form face. Twin walings allow variation in tie positions in the succeeding operations, and enable timbers to be used without being drilled and thereby weakened for subsequent usage. The lower waling must always be arranged to bring direct support to the sheathing at the kicker level and reduce grout loss to a minimum. Small cleats that project from the soldiers of individual panels support walings while tie rods or bolts are being inserted. Moulded plastics concrete spacer blocks which fit over the through bolts will, if allowed by the engineer, make for easier removal of the ties at time of striking.

The buried anchor, coil tie and spacer cone method of fixing walls becomes particularly useful at this stage. With she-bolt systems, both sets of forms can be erected and temporarily propped into position. A complete she-bolt assembly can now be passed right through the pair of forms, while the backing and waling and plated nut can be run into position. She-bolts reduce the erection operation to the stage where the only tools needed to fit the end of the bolt are a hammer and a spanner.

Mention must be made of snap ties in that these ties must be selected in the light of the design arrangement with regard to the form striking. Snap ties provide a speedy connection in the case of proprietary systems. The standard break back ties are designed for use where the tie is achieved in the plane of the form panel and where the panels are handled individually. Extended or deep break back ties should be used where the tie is made to the waling members behind the plane of the sheathing.

With single lift castings, the upper tie can be placed above the concrete thus reducing the number of rods or bolts which have to be withdrawn from the concrete. All through rods which are used as ties must be sleeved with concrete spacers, spirally wound tube plastics or paper wrapping. Whatever method is used the tie rods should be eased a few hours after casting to ensure that they will withdraw freely from the concrete. Where through-ties are used to form a thickness of concrete greater than 1 m it is advisable to substitute bright steel ties for the normal black steel, and to increase the diameter to obviate bending with consequent difficulty in withdrawal.

Tie rods of 12 mm diameter will usually suffice to resist bending during the casting of lifts or bays of normal thicknesses. Standardisation of tie sizes prevents difficulties in the maintenance of suitable nuts and washer stocks on site. The ties adjacent to the previously cast section of walling (for horizontal work) should be as close to the joint face as possible to ensure that the formwork carcass exerts a clamping action on the soldier adjacent to the construction joint, and thus to the overlapping sheathing. The actual overlap of sheathing should be kept within a maximum of 150 mm if possible to avoid grout leakage and the formation of steps due to imperfect fitting of the form face to the previously cast concrete face. Formation of a ruled joint by means of an insert fillet greatly improves the appearance of the finished concrete face at both horizontal and vertical joints.

Continuity of walings must be maintained to ensure accurate line of face and with long runs of walling the ties should be staggered.

FIG. 10.8. Special steel formwork used in dock construction. Note access for concreting operations.

Fig. 10.9. Special steel formwork which has been struck and re-erected. Attention must be paid to the corbels and overhand details, and the formwork must be detailed to allow easy removal from between the parallel faces.

This allows the walings to pass, and so obtain their support from common ties. Where returns have to be cast or where nibs forming kickers to transverse walls are required, the panels which are used to form such nibs should lie within the plane of the soldiers and should allow the walings to run continuously past or project beyond such formations. Waling material is of large section and will prove to be expensive if cut and reduced in length to overcome such problems; in fact this factor commonly promotes the adoption of a steel channel box and tube for the waling materials. Where returns and nibs are formed in this way stopend boards may be nailed between the horizontal steel to the cheek framing of the wall panel. With walling, as with most formwork operations, component parts should be allowed to oversail the sheathing in order to avoid cutting waste, in trimming to length. Whatever method is used to place concrete, on one side at least, soldiers may be allowed to project above the level of the top of the lift, since in this way they can often be used to support sheathing members and chutes which aid in placing concrete into confined or narrow spaces.

There is much to be said for framing form materials into modular panels based on board and sheet sizes. Where there is no crane available to handle the panels they should be limited to sizes which can be handled by two men, ladder fashion. For this technique the bottom of the panel is laid against the kicker and the panel run up hand over hand and linked to its neighbour by bolt or clip. After this, walers can be laid on, tie rods inserted and the wall plumbed and lined by strutting against the walings. Struts taken from the dig must bear on a plate laid on the earth and struts from a lower floor slab must be taken from plates suitably braced back from the opposing walls or from the bottom of columns; previously positioned anchors, dowels or blocks in the slab all form useful means of obtaining purchase for support.

Door openings can be formed by means of timbers punched between the wall kickers and suitably lapped at the head to facilitate striking. Window-type openings or access panel openings present problems as it is difficult to ensure compaction of the concrete under the cill member of the former. It is therefore advisable to leave an opening in one side form to allow vibration at this point, the cill member of the former arrangement being inserted and secured when the actual concrete reaches cill level. Where formers are to remain in place for some time it is useful if a gasket is incorporated to allow for shrinkage of the concrete, since this avoids tension cracking and rupture.

At one time great emphasis was laid on the provision of clean-out traps in form arrangements, but today it is generally accepted that a stopend or form panel can be omitted until the form is blown or washed out to clear shavings and other foreign matter from within the formwork. There are also several commercial vacuum lifts now available for this purpose, the only requirement being an attachment to a compressor line. Where, due to the specification which governs the placing of concrete, by chuting or dropping, concreting access has to be obtained at several points in the height of a wall form, one face of the form should be continuous and should be strutted to the full height. Suitable panels on the opposite side can be inserted and walings added as the concreting proceeds, so that the vertical alignment of the wall faces can be maintained. Where piers are formed in the walling it is useful to allow the pier forms to extend the full height of the lift, with infill panels provided between. The pier forms, which can be of a box formation, can be readily strutted and with the inclusion of extra ties can thus provide a rigidly based structure. While the tie positions are set out it is helpful to bear in mind the subsequent flooring operation or parapet formation, and to ensure that through-holes are placed and that they become available for use in providing support for these operations.

FIG. 10.10. This illustrates the nature of the formwork required where through-ties have been excluded due to the specification that governs the contract.

Between all piers or between all returns or features likely to trap forms, tapered wedges need to be inserted. Keys or other means of providing a loose section of sheathing material also allow the form panels abutting the returns to be removed from between the return faces on striking without damage on the formwork or final face. These arrangements can be removed prior to the striking operation of the main form panels, or they may be arranged to remain in contact with the concrete while the main items of the form panels are removed. In this case they can be bolted or have a dowel-pin fixing to the form and must have a splay cut lead consistent with the depth of the form at the junction with the return face.

10.7 SINGLE-SIDED WALLING

Where walling is formed against brick with asphalt tanking, or sheet piling which is to be subsequently withdrawn or diaphragm walling, it is termed single-sided walling. Single-sided walling can present problems with regard to strutting unless cast in comparatively small lifts at one time. The main difficulty is that of propping. With double-sided wall formwork, although the outside systems of supports are always required to effect final plumbing of the wall forms, these props whether tubular steel or timber do not support the actual thrust of the concrete mass, this being taken by the tie system. When single-sided walling is being cast, however, the props are required to withstand extreme pressures and though it is generally a straightforward matter to provide adequate propping at levels up to about 2 m from the kicker level, above this level there is a tendency for the tops of props to lift as pressure develops on the form face. Although such lifting action can be overcome, to a certain extent, by anchoring the form panels to the kicker using buried anchors, the resultant assembly could be an unsatisfactory arrangement which may be liable to movement as the concreting operations proceed. Once there has been any movement of such a form it becomes impossible to realign the panels, and hacking and dressing of the face may become necessary. It will be quite satisfactory to cast concrete in lifts up to 2 m deep and certainly not more than 3 m deep in single-sided work if this can be arranged. A sound method of approach is to use the 'jumbo' or twin cantilever soldier method, which employs long twin bearers. The process begins with the formation of a 150–200 mm kicker with buried anchorages at soldier positions. In single-sided formwork techniques the form operation always succeeds the steelfixing one, with the erection of the sheathing units of proprietary steel plate forms, framed timber faced or ply sheathed panels. The twin soldier members are then bolted at kicker level by bolts passed into the pre-set anchorage. Suitably spaced bolts are inserted to position anchorages in the top of each concrete lift to retain the soldiers for the next lift. Spacer blocks are required at the top of the form between the soldiers and the back face of the piles or brickwork, and it is advisable to keep these just undersize to allow for some slight deflection which will inevitably occur during concreting. Joints and connections tighten in the framing arrangement as the form is loaded and there will be some deflection of the soldier members.

When the first lift has been completed at the specified time of striking, the forms are removed and lifted to a position where the long twin soldier members pass back over the first lift to provide a

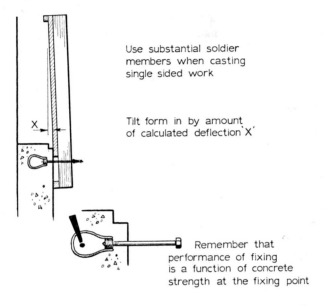

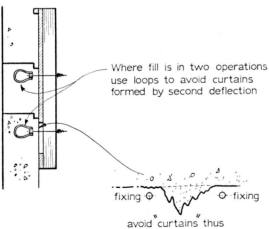

Fig. 10.11. Single-sided construction.

cantilever support. Bolts are entered into the anchorages and wedges and blocks or jacks are inserted at the bottom of the soldiers to align the top edge. A straight fillet inset into the top of the form lip gives a ruled joint to the junction between lifts. Single-sided formwork is by no means a difficult operation, but care in alignment and provision of substantial cantilever members are essential for the work to be successful.

Every consideration must be given to the sequence of operations when anchor positions are being planned. It is necessary on some occasions to partially concrete a lift prior to the insertion of starter bars or to use one form as a continuous sheathing member while lifting the form to the opposing wall in stages. Forms which are only partially filled will tend to deflect. When the lift has been completed, particularly in the case of single-sided work, a further deflection will occur, and this allows the sheathing to move away from the concrete face, to form a nib or projection at the junction between the two layers of concrete and runs or 'curtains' down the face. It is essential therefore that an anchorage be placed to maintain intimate contact between the sheathing face and the first stage concrete since this is necessary to control grout and fines infiltration.

For complicated walling operations such as those involving full storey forms to flues and lift enclosures where multiple re-use can be obtained, it is advisable to fabricate a complete set of forms, since these provide the correct profile. Cleats fixed to one face position the door openings and such like, and the whole assembly provides a working template for the steelfixer. While it is an advantage to prefabricate forms for these operations away from the working area where carpenters can work unhampered by other trades, this is not essential and many contractors combine together the first form erection and construction stages.

10.8 STAIRS AND LANDINGS

For stairs and landings which occur around central lift enclosures, it may prove helpful to obtain the engineer's permission to insert starter bars in the walling, or to provide rebates for precast landings and precast stair units. This prevents a serious bottleneck which tends to form on all multi-storey buildings where, for instance on urban work the floor areas are not large and the flooring operations can be

Construction of in situ Work 1—Foundations and Wall Formation 131

carried out at a faster rate than those necessary in forming the lift wells and surrounding work. Provided the engineers are agreeable, rebates can be formed to provide seatings for precast stairs erected

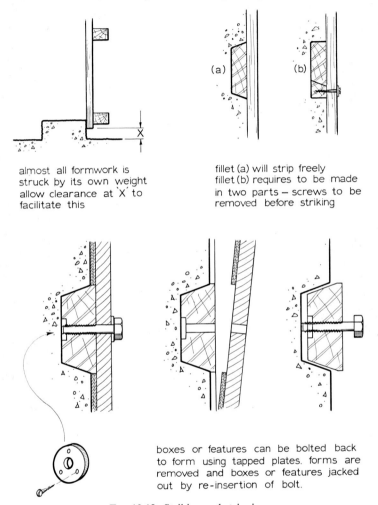

almost all formwork is struck by its own weight allow clearance at 'X' to facilitate this

fillet (a) will strip freely
fillet (b) requires to be made in two parts – screws to be removed before striking

boxes or features can be bolted back to form using tapped plates. forms are removed and boxes or features jacked out by re-insertion of bolt.

FIG. 10.12. Striking and stripping.

at a later date. Where *in situ* stairs are to be cast later, starter bars can be set aside to allow the main wall forms to be erected. Inside such enclosures corner units should be constructed which act as guides for the inner forms which give a rigid frame and which allow panels to be raised as required. By virtue of this, concrete can be placed and compacted satisfactorily within the more complicated sections of the work, since the forms can be reached by skip, unimpeded by cross-bracing or strutting. Where intermediate panels are split horizontally, the upper panels can be inserted and strutted in preparation for the completion of concreting operation, when the concrete rises to within a few inches of the joint.

If stairs are to be constructed at a later stage, casting of them must not be allowed to lag far behind the main walling operations. Should delays occur, difficulties will be experienced in obtaining access for placing the concrete because of the interference of scaffolding, projecting starter bars or props used to shore newly cast floors as the work proceeds above. Spandrel walling and stair casting can, however, produce useful work in that it can be carried out during bad weather or where it is necessary to provide alternative labour during a flat spell in the cycle of operations. Lift walling and landing slab operations require careful co-ordination. Concreting must be so arranged as to allow the casting and setting out of the kicker formation for the succeeding lifts of wall. Many hundreds of yards of decking can be laid, the steel fixed and the concreting carried out while the few operatives that can work unhindered in the confined spaces of the staircase areas can erect and concrete their own particular part of the contract. The speedy construction of multi-storey buildings lies in the employment of experienced labour for accurate handling and sound construction of lift and stair enclosures and staircase forms. This ensures productive work for the operatives who are engaged on column, slab, beam and infill work.

10.9 WALL CONSTRUCTION—GENERAL DETAILS

Cross-wall construction in high rise blocks demands the provision of unit form systems, whether they be prefabricated or built during the actual erection. Generally, they need to be substantially constructed to enable crane handling from floor to floor. Simple through-tie rods can be used effectively, with concrete spacer blocks between form faces and the spacing walls ensuring easy withdrawal of the ties. Propping and strutting for the plumbing of these walls presents little difficulty, and end walls can readily be plumbed by using a push-pull arrangement of telescopic props which bear from and tie to a previously cast bay of a floor. Reinforcing steelwork for extremely tall slender structures may be complex. Bars cut to long lengths which span more than one storey of a wall need to have a supporting scaffold for the steelfixers' use. Where this is required it is essential to have a close co-operation between the various trades to avoid form panels being tied up where the crane cannot pick them for transfer to the appropriate positions for the next operation. Scaffolders have to switch the outriggers that support such steelwork from tower to tower in order to release form panels as and when they are required for re-use.

Where comparatively shallow lifts of concrete are repeatedly cast as on towers, storage silos, chimneys and similar structures, a variety of panel formations can be adopted. Generally, the formwork consists of a sheathing in steel or timber, or composite construction, which is supported by vertical bearers on straight work, and horizontal sweeps or bends for curved work. With plane faces the direction of the boarding, if this is used for sheathing, is horizontal, while for curved work such as that for chimneys, towers and settling tanks, boarding or lagging will lend itself to the shape more easily if run vertically to span over shaped ribs or bearers. For a plain concrete wall, formed by climbing forms, much of the work is repetitious, but once the methods of handling and positioning ties have been established, it can be used to provide a regulator or tempo-setting part of the work.

When the plan of a structure is of a regular nature, panels can be prefabricated on a production basis and delivered to site ready for erection. The major tie holes can be pre-drilled and provision can be made for the connections between the panels to be effected by the insertion of bolts or wedges. If proprietary wall clamps are to be used, either the scissor type or the infinitely adjustable type, these can be fixed to the panels to provide lifting points for transportation by crane or chain block from level to level. Where timber soldiers are used, loops or rods of reinforcing material, inserted through holes bored in the tops of the bearers when used in conjunction with a strong beam or spreader, provide a satisfactory way of facilitating the hoisting of units. An efficient way to lift unit formwork is to provide outrigger brackets from the scaffold tower, or bridging tubes from the inside to the outside scaffold; this provides quick action, portable hand winches of the type used in warehousing being available. These can be hooked over the appropriate bracket or tube, and the unit can be lifted quickly into its new position, the operation being repeated with every panel along the length of the form arrangement.

There is little doubt that it is advantageous to use chain blocks to raise sections of wall formwork on work that has to be carried up to any great height, where a tower crane is not being used or where weather conditions are likely to be bad. Provided care is taken over the positioning of the outriggers, the gang responsible for this operation have their lifting arrangements within their own control and can work independently of the crane schedules for other operations. The forms can be hoisted clear of the concrete for cleaning, and access provided for possible modification by the insertion of packs and pads. Inside panels can be hoisted away from the work when, for instance, floor casting is taking place and the forms do not take up valuable space on the working platforms, thus causing hindrance to associated trades between uses.

Simple tie rods can be used for retaining formwork when chain blocks are used for hoisting operations. Through holes can be avoided if the bottom tie is positioned over the top of the previously cast lift. It is thus possible to keep the upper tie out of the concrete lift, since a saving in time results which would otherwise be spent in withdrawing ties from the concrete and re-entering them to provide bearings for the panels as lifted. When loose panels are retained by through-ties, and are hoisted by crane or by some other manual technique, support is necessary once the lifting action has been completed.

It is dangerous to rest free forms on a tie which passes over concrete because there is no positive support for the shutter until both the opposing panels are erected, and the bolts are tightened. Where plain through-ties are being used with climbing forms, it is necessary to insert a number of the top ties through the concrete to provide support for the form panels while the remainder of the ties are being inserted.

When forms are lifted manually or by crane, there is an immediate demand for the facilities which are offered by the buried anchorage, or coil type of tie in that one of a pair of forms may be unbolted and its opposing panel can remain securely fastened in position, so that cleaning and the application of the parting agent can be carried out before the panel is repositioned after the steelfixing has been completed. Panels which are not being moved are always completely safe, and individual sections of panels can remain bolted back to the concrete until required for re-use.

10.10 CIRCULAR AND CONICAL WALLING

When boards are used for circular work, according to the specification, the board width needs to be selected to form the permissible size of facet. For high quality faced concrete (using board sheathing) it is necessary to mould the surface of the board and bevel the edges of the boards in order to produce a close joint. Close joints occur on inside of forms where a V-shaped groove would otherwise occur, and be most marked on work which utilises smaller radii. Multiple layers of plywood of an appropriate thickness to the radius being formed or for larger radii present a jointless face. The adoption of layers of plywood effects savings in machining which would otherwise be required to present a comparable finish utilising individual boards. These savings become extensive where forms to the conical faces of settling tanks and hoppers are being used, although the wastage factor involved may be slightly increased beyond that for plane or circular work as developed panels need to be cut from the ply sheet.

In principle, the formation of circular or conical forms for walls is similar to that adopted for plane forms. The shaped members may be bandsawn and moulded to the required profile, and the ribs can be spaced by rails at the edges of the panel if required. Edge rails become necessary on conical panels where there is a tendency for the bevelled ribs to turn about their axes when the panel is loaded. Where panels for conical work are being formed on timber backings and the services of a woodworking mill are available, it is advisable to provide the mill with a template of one of each of the panel ribs to be used while the material is bandsawn to profile. This template can, where the quality of form being built so warrants, be used against a collar on the spindle moulder to provide a machined face suitable for glueing or achieving the fixing for the sheathing face. This ensures that moulded boards or the back face of ply sheathing will bed down and present an accurate profile at the concrete face.

For circular and conical work, such as that involving plain work between reveals, tapered wedges or splayed fillet should be provided within a complete ring of forms to assist in striking. These keys, or striking fillets, are either bolted or dowel-pinned into position while the concreting is being executed. It is essential on conical work to provide for strutting of the outer face in order to support the vertical

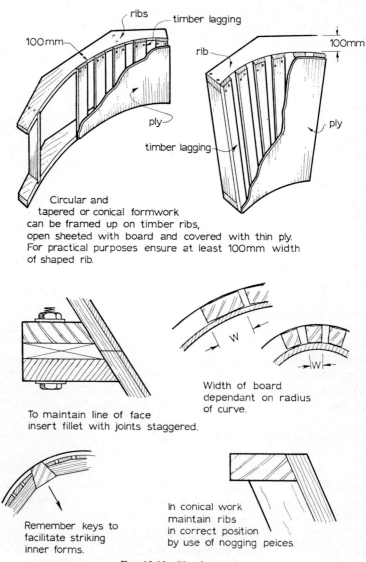

FIG. 10.13. Circular work.

loading which results from both dead and live loads. At the junctions of the rings formed by the shaped ribs, intermediate rings of shaped timbers should be inserted to retain the form panels to the true profile. Alternatively, the panels in the adjacent rounds of form panels may have their vertical joints staggered to bond the succeeding rings of panels, and to retain the correct face profile.

Depending on the properties of the concrete mix, the inner faces of hoppers are screeded or formed to the eventual profile. It is preferable on conical work to employ forms similar to those that are employed for the outer face. These panels are fixed between braced runners which span from the hopper bottom as far as possible up the wall in the concreting direction, the runners being spaced from the outer form to ensure the correct thickness of walling that is being cast.

Other types of walling form encountered on *in situ* concrete work are those used to form tapered walling to the head of hoppers, the bull-nosed corners in retaining walls and similar taper induced arrangements. To form this type of work the principle adopted for the cooling tower shell walling can be used. The maximum variation between the width of a filler board, between panels at the foot of such a wall, and the width of a board in the same plan position at the head can be calculated. This difference is the necessary total width of packers which needs to be removed from the make-up or infill board as the form progresses from the bottom to the top of the wall to generate the taper. If the difference is divided by the number of lifts in the process of moving the form over the lift height, then the resulting dimensions are the actual dimensions of the slivers or pack pieces. The difference over a number of uses may be such that it is advisable to remove the initial make-up board or pack, and substitute a pack constructed of similar elements but with the main pack members being of reduced width. In this way extremely fragile packs composed of a multitude of slivers can be avoided.

Where panels are required to form parallel lifts of circular walling, followed by lifts which taper in profile or section, the initial make-up packs are parallel and, at the point of change of section, these can be replaced by a tapered pack, the overall dimension of the widest part of the first tapered pack being equal in width to the initial parallel packing piece. Tapering walls can be formed in a similar way with one or two faces made adjustable by the insertion of packs.

It is necessary when taking off ironmongery for use with such formwork to allow a sufficient thread on the tie bolts so that they can be used for the required number of operations without any need for replacement. For buried anchors, the appropriate lengths of tie are ordered so that they can cope with the various thicknesses of wall on the succeeding lifts. Where she-bolts are used the studs vary in length.

Timber forms for circular and conical work are generally purpose made for the particular radii involved, and thus demand individual setting out of all ribs and shaped members. It is possible, however, using ply as the sheathing material or board panels that can be flexed to shape, to use purpose-made rollings or steel walings to cover a variety of radii by the insertion of long tapered wedges between the sheathing and the wall. Obviously this is not an economical proposition between individual lifts of concrete. Where there are a reasonable number of uses on parts of one contract, which differ only in radius from further uses, then it has proved worthwhile to adopt this method of springing forms from a rigid backing member. The backing member can, if necessary, be re-rolled or recovered once the casting is completed for further use on other contracts.

The contemporary architect often includes spandrel walls between columns at the perimeter of a building, and these can either be flat or curved in plan. Such walls are formed in any of the ways previously described, on to an accurate kicker formed during the concreting of the slab proper. Another provision, that of ribs on columns for receiving the ends of spandrel forms or buried anchors in the column face, helps with the erection of walls when carried out as an operation subsequent to the casting of the columns. With columns that spring from the spandrel the kickers for the upper part of the columns should be cast with the spandrel waling. Column bracing becomes an easy operation if the anchorages are cast in, to allow the fixing of continuous runner supports for the back or outer form faces of the columns, so that a satisfactory line of component is achieved.

Walling, in its various forms, represents a major section of the concrete work. Generally, progress governs speed of execution of the contract, and often reflects the amount of care applied to the design and provision of formwork. Attention to the details of ruled joints go a long way in enhancing the appearance of the completed work. The whole process is very much one of mechanical handling and, provided the applications of the skills of the method engineer and planning engineer are directed properly, much can be achieved in producing a good job. Where there is sufficient re-use to warrant the adoption of a traveller, this serves to uncouple the walling operations from the handling operations normally adopted on site, and improves the productivity of the walling team.

Chapter 11

CONSTRUCTION OF *IN SITU* WORK 2—COLUMN, BEAM AND SLAB FORMATION

11.1 COLUMN FORMS

The provision of column moulds is an important part of the formwork programme of most contracts. Construction varies according to the materials used and these are mainly decided by the specification. Probably in present-day building work the great proportion of reinforced concrete columns and concrete casings to structural steelwork are cast in accordance with the specification for sawn formwork where the columns and casings are to be plastered or decorated. Many contractors use fine sawn or wrot boards cleated together by offcuts of the same material, nailed and clenched in position. These panels are retained by tie rods through the cleats on one set of opposing sides, while inserted wedges are used to support the other sides by wedging from the ties. Alternatively, it is more efficient to use column clamps in sets which encircle the form with some sort of wedging adjustment. Banding and strapping are also used in place of traditional ties. For simplicity of form construction ply sheathing on timber backings is frequently adopted, particularly where the specification calls for smooth concrete of consistent appearance, or for the use of wrot formwork,. Plywood panels require backings which are consistent with the thickness of material used and the head of concrete to be retained. Care should be taken with the method of lapping the backing member at the corners to afford positive register of the panel edges and accurate sizing of the column. Correctly interlocked forms prevent grout loss and honey-combing of the arrises. Columns over 600 mm square in plan are best cast from ply-faced forms suitably framed with 100×50 mm, or similar section, backing material as used for wall-panel units. Where

Construction of in situ Work 2—Column, Beam and Slab Formation 141

large sections of concrete are cast, such as heavy foundation columns and columns that support the lower floors in multi-storey work, it is advisable to dispense with column clamps and use twin walings with tie rods which pass and stagger on adjacent sides of the column

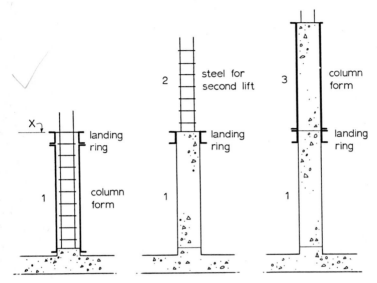

1st operation column form and landing ring erected around reinforcement column cast to level X

2nd operation landing ring remains, form removed - steel fixed

3rd operation column form re-erected onto landing ring and column concreted

Landing ring provides for accuracy of line and supports forms in tall column construction.

A similar technique can be used to support wall form panels.

FIG. 11.1. Column construction.

assembly. Here the lower pairs of walings are placed close to the line of the top of the kicker to ensure that the sheathing is securely clamped back at this vital point in the structure.

All column forms, where there is a great proportion of reinforcing steel or where the casting height is extreme, should be so constructed and clamped as to resist the rigours of concreting and compaction. Considerable effort is usually applied to ensure that compaction is achieved efficiently in the lower part of the form. Columns are often constructed with opposing panels in two heights to allow placing and compaction. The formation of columns which have nibs or brackets for intersecting beams may present certain problems with regard to the sectionalisation of panels. These problems can be overcome by the judicial jointing of the form sides and by the insertion of stripping fillets at points where binding may take place. Provided the engineer agrees, where beam systems intersect at the column head, concrete may be stopped off at the level of the lowest beam soffit, the head of the column being cast with the beams and slab. Quite often a considerable amount of time and trouble has to be spent in the grinding and dressing of arrises for columns and beams where the corners have been damaged during the striking of the column forms, or where honeycombing due to grout loss is apparent. Care in forming the corner joint, and the use of a built-in feature fillet instead of planted fillets, will help to overcome these difficulties. A great number of badly formed chamfered corners are caused through the infiltration of grout behind say 25 mm × 25 mm planted fillets where specified. With vibration, grout infiltrates into any joint which is not kept tight by clamping action. The junction of the splay with its feathered edge at the panel face when formed in the conventional way offers a direct route for the grout to infiltrate under pressure. A supply of suitably sized materials which can be built into the corners of such moulds, and also where bull-nosed corners are specified ensure that the panels can be constructed such that the action of clamping provides tight butt joints into which grout will not enter during concreting. In practice, such a fillet will only slightly increase the cost of materials. The dressing work can be considerably reduced and a saving in labour which would otherwise be expended in the replacement of the plain chamfer fillet, since the plain fillet is often wrenched from the form face by the grout fins during the striking operations. Where steel column guards are being cast into the concrete, it is necessary to have tapped holes, previously prepared in the guards to enable them to be bolted back to the form panels to avoid displacement by vibration during the actual casting. The retaining bolts must however be withdrawn prior to striking. The top and edges can be sealed against seepage by plastics stopping, or foam strip.

The column-forming operations must begin with the casting of an accurately placed kicker centred off from the starter bars, and positioned in relation to the lines scribed, or mortar dots laid, during the setting out procedure of the foreman or engineer. For large columns, buried anchors may be inset at the centre line of the column to retain the form face in close contact with the kicker. Form panels which are brought to a suitable width by packers or rips are offered into position, and the joints are spiked temporarily to locate the individual panels while the clamps or bearers are attached and tightened by wedging or bolting. The arrangement is then plumbed and lined by raking struts which take the form of adjustable props that bear against the foot of other columns or on to plates on the concrete slab from which the columns spring. Columns are concreted to the full height, in the case of slender columns, and to the point where the upper panels are inserted and cramped into place on large columns. As the concreting proceeds the columns must be checked for plumb, and any slight adjustment made as required.

Splayed columns, where either all faces or only the opposing faces are splayed, require special treatment. Depending on the degree of the splay the bearers may need to have a bevelled edge or face. Where only one face is splayed, as often occurs with portal column sections, the panel which forms the splayed face can be of net width and fitted between the cheek panels which will be the wider pair of the four panels. The face panel may thus be retained at the appropriate pitch by bearers which are screwed or bolted within the face of the cheek forms. All tying and strutting can be carried out in the normal manner employed for square forms. Where two faces are splayed this treatment can be carried out to both opposing faces. Where a break or offset occurs in the splayed face it is quite a simple matter to set out a fillet on the cheek, and to allow for the thickness of the sheathing and bearers or runners. Columns splayed on all four faces are treated in much the same way, although the edges of the internal forms need to be slightly bevelled or backed off to produce a good, grout-tight fit at the corner and thus preventing the loss of fines. Columns that form dormer cheeks in mansard construction can also be formed in a similar way. When such columns call for a removable panel for concreting purposes, it is a simple matter to form a break in the inner form face and slide in the panel so formed between the cheeks and against the retaining fillets.

Contemporary architecture often demands mullion type columns cast at close centres to the perimeter of a floor area. With such slender work it is essential that the columns be perfectly plumb and in line, the slightest dimensional error being immediately apparent to the eye. It is a good plan to construct the outer face panel of such forms on a continuous framed grillage which spans across four or five columns in the line. An accurately constructed grillage of this nature, properly braced and framed by screwing, requires levelling only in one direction, and all columns can be accurately spaced and plumbed relative to the elevation of the building. The remainder of the panels which complete the individual 'boxes' can be erected using the normal methods as adopted for individual columns. After plumbing, the columns are cast and accurately spaced for window fixing or subsequent panelling operations.

Circular, elliptical and irregular shaped columns require particular attention at the design stage. Unfortunately it is usually the case that few such columns are placed for architectural purposes at key points in a building. With little re-use value obtained from the forms constructed for casting, a very economic method is demanded. Where, as is often the case, only one circular column is to be cast, it is worth enquiring for appropriate sized cast asbestos or plastics pipes. These pipes can be supported by runners which are wired or suitably tied to provide lateral support; the pipe can be broken away after the concrete has cured. A form constructed from pipe may be used more than once if split on the centre line and framed up as for timber sheathing. Proprietary foil and resin impregnated tube forms provide economical methods of forming circular columns where it is not possible to fabricate special re-use panels. Circular columns fabricated in moulds of glass-reinforced plastics have recently been cast using a method where a cylindrical mould with minimal draw has been struck by lifting off the newly cast concrete without dismantling.

For elliptical and irregular shaped columns, it is necessary to adopt the traditional methods of the form builder using a form which consisted of suitably sized lagging members screwed or nailed to shaped diaphragms, which in turn were supported by runners and clamps or bolts. Prior to the availability of reliable plywood for work in exposed situations, it was necessary to use segments sawn from timber for the yokes. These required the manufacture of the forms in four sections to allow the segments to be cut from economic sections of timber. However, as resin bonded plywood is now available it is possible to manufacture such forms with ply yokes in two panels only, the whole being simpler and more economical to erect and plumb.

Circular, mushroom-headed columns are virtually the province of the steel and glass-reinforced plastics form manufacturer, although of course they can be cast using timber formwork. With such columns the contractor would do well to approach the engineer with a view to obtaining his permission to cast the columns in two operations. The first operation involves the plain circular column casting, and the second the casting of the hopper head. Square columns with splayed heads do not present so much problem but may be cast in this manner if allowed. Failing this the hopper head panels can be framed on to the plane face of the column, but the construction would then involve framed cleats bolted at the corners to act as ties where the splayed head size, combined with the form thickness, precludes the use of clamps to retain the assembly. Great care must be taken to tighten these bolts to avoid the joints opening during the concreting operation. Applications such as these mean that packing-case strapping, linked by patent crimping or clamping machines, can be extremely effective. The banding and clamping tool may either be purchased or, as is often the situation, hired for a nominal sum. The actions of clamping and tying are combined by two actions of the machine: jacking to tighten and crimping or clipping, by activating another lever. The system is found to work most efficiently when the corners around which the banding passes are slightly rounded, and on wide forms when intermediate blocks are inserted. The inserts form a further series of pressure points beside the four corners of a column or beam form where the banding has its most intimate contact. Banding has been used successfully on forms which have shaped bearers to maintain contact between the bearer and the band. On the larger pier-type formations, however, the twin yoke and tie rod or heavy column clamp methods of tying prove to be more positive.

One advantage of the steel strapping method of form tying is that there is very little waste of the tie material. The pieces may be joined and rejoined, and provided the clips or joints do not coincide with a corner of the form to become distorted and thus fail, such joints have the full mechanical properties of the unjointed banding material. In certain instances greater economy can be achieved by allowing some spare strapping to extend beyond the fastening for use in subsequent operations. Striking formwork so tied becomes a speedy operation when carried out with snips. Many methods of tie formation have been adopted over the years in an attempt to reduce the erection time of column formwork to a minimum, but although certain methods would appear to be advantageous, the simple methods which utilise tie rods or proprietary clamps invariably prove to be more economical over the contract period. Properly maintained ties and clamps offer continuous re-use value, and with captive nuts and wedges are not easily mislaid and thus possibly cleared from site with rubble, as sometimes happens with parts of the more complicated systems. Probably the greatest time savings can be achieved by reducing the number of components to be handled. L-shaped forms, and even double L-shaped ones, reduce the components in simple square, rectangular and paired column casting.

While on the subject of column casting, it is useful to consider the methods of locating brick ties and slots for fixing stone facings and the like. To fix back wire ties for infill brick panels in any great quantity, impact staplers which use long staples provide a rapid fixture, since a skilled operator can work at twice the speed of one who uses a hammer and nails. For the fixing of a masonry channel where there is multiple re-use, rubber blocks screwed to the face of the form provide a press-fit location of the channels and soon repay their initial cost through the speed of the operation. In every case the masonry slot must be taped to exclude grout.

Earlier mention was made of the method of casting-on column caps to the upper part of a column, where the beams intersect and it is advisable at the time of erection of such cappings, and indeed on succeeding lifts or where tall columns are involved to clamp a dummy yoke around the column direct on to the concrete to afford a 'land' for positioning the individual panel sections of the cap form. In civil engineering terms these yokes are known as landing rings and may comprise a band of formwork attached to the top of a lift of panels. On work which involves the varying of lift levels where the concrete has been cast to the level of the soffit of the lowest intersecting beam, the yoke can be spaced some 150 mm below this point, the capping arrangement rests on this and can be clamped back by another yoke or by bearers which are integral with the form panels. Where the beams are of a regular formation throughout the various levels of the building, the upper part of the column may well be formed by cheek pieces attached to the side forms of the beam. The purpose of the column capping form is generally to overcome variations in the sizes of the intersecting beam. Where the cheeks are attached to the form sides, the yoke fulfills the dual purpose of supporting soffit ends which abutt the column and the cheeks of the beam sides.

On the question of column formwork, mention should be made of the linked columns that often occur at the corners of cold stores and framed buildings which have differential movement joints. Several economies can be effected by careful planning when deciding on the panel arrangement for forming such columns. Where the two parts of a linked column have similar profiles, one set of forms can be provided which are capable of modification to both outlines. By this technique the end panel from the first operation can be removed and the form inverted, if required, to cast a handed version of the original profile. To facilitate the inversion, it is necessary to keep the yokes or bearers spaced at regular centres. During the casting of the first column of a pair, through holes or inset buried anchorages have to be formed to allow the form panels to be tied back during the second casting operation. *In situ* columns, which possess through holes for services and seatings for precast beam systems, present similar concreting problems as do walls where window openings provide difficulties in achieving a satisfactory consolidation of the mix. Care is required to avoid the formation of cavities, due to trapped air, below the openings. Where large through holes are to be formed in columns, it is desirable that the column be cast to the level of the bottom of the opening, and then a panel inserted which forms the bottom face of the opening. This panel should be securely fixed to the form side to avoid the concrete surge within the twin section of the column, which could lift the panel and so produce an irregular face at the bottom of the opening. When large openings have to be cast for services, it is advisable to provide openings in the form sides so that a number of formers can pass right through both panel cheeks. By adopting this method the steel reinforcement cage may be erected in the normal fashion without any interference from the forms.

Columns which have nibs that project for supporting precast units, or which form the bearing for a crane rail, will require similar treatment. The concrete, in these instances, is cast to the nib level, and cover boards are added while concreting is completed to the remaining part of the vertical section of the column form. It may well be that the engineer will allow the main section of column to be cast with perhaps a joggle or recess being formed, while the nib is concreted into position as a secondary operation. Here the form panels are slotted to allow the reinforcement to project while the column is being cast. This method of casting will again necessitate that dowel bar holes or buried anchorages be let into the concrete to facilitate the fixing back of the corbel form in the secondary operation. Continuous nibs to one or more faces of a concrete column may be formed by stooling out the form face to provide a recess of the correct profile. Care must be taken to ensure that there is no grout seepage at the joints between the stooling and the form face. Where this stooling is likely to become cumbersome and difficult to retain in its correct position within the form, it will be more effective to construct the actual form face to the profile, so that good use can be made of the twin-bearer method of tie accommodation generally used to avoid boring of bearers for tie rods, and to eliminate the need for unthreading nuts and washers from ties while the form is being erected. Light framing members can be inserted between these twin bearers and used to form a shaped bearer arrangement to which the sheathing boards can be fixed thus providing the required column profile. This method is particularly effective with large irregular shaped columns, and columns which are situated adjacent to entrances where cheeks for canopies or decorative features are to be cast.

When forming columns for multi-storey buildings it is necessary to establish an optimum section within which provisions can be made for varying sections or profiles. The sections should be grouped from the schedule into types, and the required number of forms or 'boxes' allocated to allow the programme to be maintained. The number of column forms to be supplied is governed by the rate of the casting of the concrete floors. Little is gained from having rows of columns which are cast too far ahead of the decking gangs. When the number and various main types of forms to be provided has been ascertained each casting operation to which the column form is to be applied should be examined. Joints of suitable width are inserted in the panels to allow adjustment with the minimum amount of carpentry work between uses. Where board is being used it is a simple matter to add rips to one edge of the form to vary the section to suit major alterations. If plywood is used it may prove more effective to hoist a light bench saw, or powered hand saw, from level to level making adjustments as the work continues. With the appropriate amount of thread allowed on tie rods, and a suitable size of clamp chosen to accommodate the largest possible number of variations in section, column-form erection becomes an almost mechanical operation with sufficient columns being cast daily to provide an area for the decking gang to follow through. Once the first bays of slab have been concreted, the kickers are set down and the column forms erected on the next level. Some contractors aim to have as many as five storeys under construction on suitably shaped sites at one time thus providing constant employment for all the trades concerned.

Where particular columns are of a complicated section such as at intersection of return walls or adjacent to staircases, a special set of forms should be provided for this use alone and not to modify those used to cast other columns. While this is perhaps contrary to the basic principle of maximum re-use it ensures that difficult applications are formed in the most efficient manner, so that valuable time is not expended on intricate modifications which could otherwise be spent on more productive items elsewhere. It may be that a form panel can be kept as a standby for use on one particular position at each floor level to avoid modifications to the standard forms used throughout the rest of the job. Time spent in providing such special panels and the extra amount of material used will soon be recovered over several uses in multi-storey work.

Where dimensions of column sections are suitably standardised throughout a building, it is economical to use steel column moulds, the contractor being well advised to maintain a number of moulds as a plant item for this purpose. Indeed with the increased cost of timber and plywood, the gap between the cost of the steel purpose-made form and the timber component is narrowing to a point where steel may prove more economical over a small number of uses.

11.2 BEAM FORMS

Following on the striking of column forms comes the erection of the beam soffits and sides, and the success of many contracts rests on the efficient design of such items. With the columns erected a grillage will be required to support the soffit bearers and sheathing. Where the columns are of small section, it is advisable to place runners against the face of the head of the column and so form continuous supports for the sheathing or bearers which can readily be lined and levelled using adjustable props. The runners can be made to pass by being staggered out of line, while props with wide head plates ensure their support without rotation under load. The use of forkhead-reveal pins within the head of the adjustable props is a good course to adopt both for rigidity of support and as a safety precaution in preventing movement during the placing and vibration operations. Forkheads, wedged where necessary, also ensure that the axial loading is applied to the tubular member. Cross members are laid directly on to the grillage formed by the runners to support the soffit sheathing, or alternatively framed panels of sheathing secured to bearers can be laid into position.

FIG. 11.2. Precast diaphragms which are used to position ducts for post tensioning tendons, are here supported on the main beam soffit which is in turn supported by a falsework of 'transportation' trestles.

For haunched beams the runners, less the thickness of the bearers and sheathing, can be cut to profile so that when these are placed and secured the desired profile is generated. The bearers should project well beyond the soffit line to form a landing for the side panels of the beam, and so to allow the fixing of a plate from which the side panels may be wedged and strutted. Side forms may be arranged to support the edge of the sheathing material to the slab by direct bearing or by provision of a fixed fillet upon which the joists to the main slab area will bear. This method calls for a clearance between the end of the joists, if in one length, and the back of the bearer or sheathing to facilitate striking. Where beams are being cast prior to the casting of the main floor slab it is possible to allow the tops

Construction of in situ Work 2—Column, Beam and Slab Formation 151

of the bearers to project and to employ tie bolts and spacers to retain the form sides. On reinforced concrete work it may well be permissible to leave through-holes in the beams for tie bolt or dowel bar insertion so that anchorages for plates which support the ends of the joists to

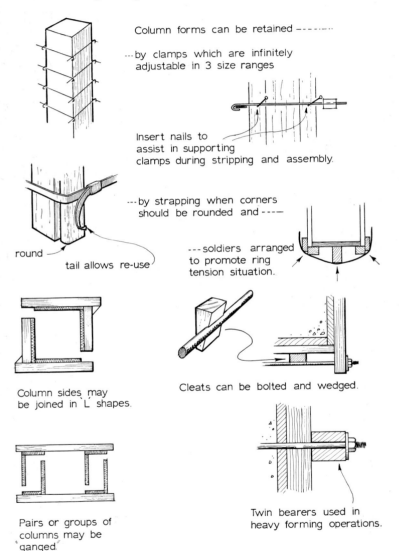

Column forms can be retained --------

--- by clamps which are infinitely adjustable in 3 size ranges

Insert nails to assist in supporting clamps during stripping and assembly.

round
tail allows re-use

--- by strapping when corners should be rounded and ----

--- soldiers arranged to promote ring tension situation.

Column sides may be joined in 'L' shapes.

Cleats can be bolted and wedged.

Pairs or groups of columns may be 'ganged.'

Twin bearers used in heavy forming operations.

FIG. 11.3. Column casting.

the floor forms may be formed; alternatively buried anchors may be inserted for this purpose.

Deep beams should be formed in exactly the same way as the walls, through-ties being desirable for all beams over 600 mm deep. Toe beams and beams with projecting nibs can be cast by stooling from the main former, and may be of such size as to warrant a built-up bearer member, which utilises the space between the twin soldiers to fix diaphragms or firrings upon which the sheathing profile is built up.

Where the perimeter beams to a floor slab are being cast monolithic with the floor it is essential that through-ties be inserted to overcome the difficulties of providing suitable raking struts to retain the outside form panel. Alternatively, a line of standard bearers, longer than normal, should be allowed to project and these can be strutted using raking props supported from tapered plates. The outside scaffold should not be used to provide purchase for raking props. Because of its design there is always some measure of movement in scaffolding due to the working load, which can result in a most unsatisfactory strutting arrangement. Proprietary beam clamps are useful where beams are of a fairly standard depth and clamps can be chosen the legs of which should not foul slab forms or undercarriage work for the casing of structural steelwork. Buried hangers provide an efficient method of support for the larger storey heights and where it is not possible to prop from floor slabs. Adjustable props and a standardised treatment of forms are economical for the majority of structural steel casing work especially where slabs are substantial and storey heights are moderate. For casing small structural steel members steel column clamps can be used to perform the function of both hangers and ties, using concrete spacers to position the formwork relative to the steel. One arm of the column clamp rests on the steelwork, while the remaining arms support the side forms and soffit.

Where there are large numbers of beams formed in a slab such as are found where heavy equipment is to be installed and the intervening spaces between the beams are utilised as ducts, a complete joist and bearer grillage should be run at the level of the soffits to the main beams. The beam sides and floor soffit forms can be supported from this, the joists and bearers which form the slab soffit acting as struts to retain the side forms. In the case of small section secondaries which span between the deeper main beams, cheek returns formed in the main beams at the intersection points locate the secondary beam moulds. Removing the ends of the secondary beam sides from the plane of the bearers to the main beams facilitates the striking of the assembly. Where permissible, wire fingers inserted through the formwork will retain the beam soffit and side forms while the supporting grillages are being struck down, and thus prevent damage to the framed or fabricated panels. The fingers can be snipped at the time of removal of the panel, and the wires which are cropped or broken off with the face of the concrete require filling when the face is dressed.

Beams circular or shaped on plan are treated in the same way as circular walls. Shaped timber ribs can be provided which are faced up with a layer of board lagging, sheathing plywood or a combination of the two materials. A ply layer provides a good wear-resistant face while intricately shaped and small radii may be readily formed using this method, which can also be used where circular ramping beams have to be formed.

Shaped ribs for ramping beams are cut, and can be clad by covering with board laggings moulded on face, assuming extremely small radii are being negotiated. Over this is placed a ply, or for low re-use work, a hardboard sheathing. Ordinates can then be set out for the soffit level and timber fillets may be screwed or bolted back to provide a bearing for the ends of bearers which run from one side form to the other. Ply bearers cut to the developed profile of the beam soffit can be laid on these. This type of construction can reduce considerably the amount of setting out required when compared with that involved in providing a framed soffit panel. The form panels, if allowed to run well below the soffit level, permit through tying below the level of the concrete.

The shaped ribs which support the side sheathing should, where the radius allows, be cut from single timbers and shaped on both edges, and if they are cut to the plan profile of the beam, they can be staggered and nailed or screwed together to form continuous ribs. Soldier bearers can be used to facilitate tying of the beam. If it is not possible, due to the radius of curvature, to shape both sides of a rib because a member may fracture along the grain, then seatings have to be cut to make allowance for the bearer fixing.

When spiral staircases are formed to a tight radius the inside string or face of the beam is best formed by erecting a complete drum to the full storey height, and by sheathing this with a band of plywood to act as the inside string, several layers of thin ply can be used to form very small radii. The drum can be collapsed and hoisted when required for further use, or for a one-off operation simply broken out and the material recovered.

11.3 STAIR FORMWORK

In situ stair construction represents a good cross-section of all the means employed in casting concrete since it is a combination of slab and beam casting. The fact that progress in the formation of stair flights, and their accompanying structural components in walls and beams greatly affects the general concreting progress on site, has been established elsewhere in this book and obviously cannot be overstressed. Unfortunately, in structures where mechanical lifts are installed, the stair formation has to be tied to the wall structure which encloses such lifts. Stairs are often complicated by the fact that they are generally made to fill any awkward spaces which result in the overall design, and therefore extreme accuracy of setting out is demanded, both in plan and elevation.

Simple stairs with straight flights offer little difficulty, the soffit being formed with steel plates or plywood. The support is taken from raking props or puncheons against plates which are either strutted back from adjacent walls or fixed into the floor slab. The stringer face is generally formed by timber panels or plywood on a framed backing, the riser boards being supported from cleats screwed or nailed to the inner face of the stringer. To help with the operation, of trowelling-off the concrete tread, the board should be backed-off so that trowelling can be carried well into the intersection between the tread and the riser.

A common means of securing string panels in position involves the use of raking struts while soffit bearers are allowed to project for this purpose. A rapid means of support is achieved using beam clamps which span the flight bearing back against the runners behind the string sheathing. Where stairs are formed adjacent to brickwork, the string can be formed by an inverted cut string panel. A substantial plate should be spiked or punched back against the wall, and adjustable props and cleats can be dropped to the appropriate position to support the riser boards. Where the materials which are needed to form soffits to landings and stairs are to be subject to multiple re-use, it may well prove expedient to frame up half landing panels. These panels can bear on plates which are attached to concrete or brickwork, while the soffits to the flights can be framed and arranged to span between such landings. Support is effected by bearing from the landing panels, and by propping or punching against the kicker or lead up section of the lower landing.

Form construction for staircases is generally not complicated in itself, but there is a need to maintain accurately finished concrete levels at all points to facilitate the floor finisher's work. Constant checking of the form details against the structural drawing, by means of a storey-rod marked with tread levels, is necessary. There is often a different thickness of finish applied to the structural floor, landing level and stair treads and it should be noted that a storey-rod will maintain these dimensions accurately. Where balusters are formed, these are cast within a solid string panel and an inverted cut string panel tied back in the conventional manner as for upstand beams. Projecting bearers, spacers and bolts provide support. The whole arrangement is then strutted by raking struts from the soffit bearers, puncheons from adjacent walling or plates between columns.

11.4 STANDING SUPPORTS

It is necessary, depending on the design of the structure, to form day joints or construction joints in beams in varying positions. Invariably the contractor is required to provide standing props to support the structure, for instance while constructional loads are imposed by work on upper floors of a building. Reshoring, as such, is not considered to be good practice by many engineers and when it is considered that several tons of force can be applied to beams in complete opposition to their designed loading by reshoring, using adjustable props, this viewpoint is understandable. It is preferable to leave standing pads on sections of the soffit, jointed in such a way that they remain in position after the main soffit members have been removed. Props set below such pads prior to concreting can remain untouched throughout the striking operation and until the concrete is capable of supporting the constructional loading and dead loading of the floors above. Care must be taken to ensure that the standing pad positions on succeeding floors are such that continuous props are formed, thus avoiding eccentric loading in the intermediate floors caused by staggered props. This precaution should be applied, particularly where there is a rapid turnover of material on a multi-storey building and where for instance there is a bay of flooring having a cantilever projection or curved perimeter beam. Standing supports overcome the tendency of the beams to rotate until sufficient strength has developed in the concrete to resist the turning moment. Cantilever

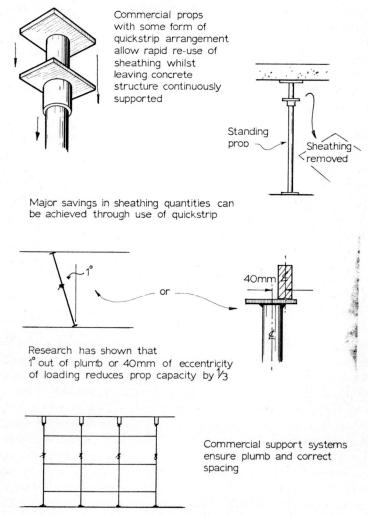

FIG. 11.4. Prop considerations.

beams have become a feature of contemporary construction and require special attention where standing props are concerned. Where possible the soffits to cantilever members should be left intact to avoid reshoring. Any raking supports which form standing props must be tied back to the columns to give a pure support. As a general rule it is advisable to avoid partial removal of the support for cantilever beams and slabs since an incorrect removal sequence can bring about failure of the structural component.

A cantilever component supported at its free end becomes what is effectively a simply supported beam. The reinforcement is mainly at the upper face of the component and there is considerable likelihood of tensile failure in the bottom of the slab or beam. Standing supports should be provided in the form of continuous strip supports which allow the concrete full support until capable of sustaining dead and live loading.

In multi-storey construction where standing supports are continuous through several floors it is advisable that the adjustable props be eased to allow the more mature lower structural components to contribute to the support of loads from above. Normal reinforced concrete beams and slabs do not sustain loads until deflections have occurred, and continuous support without easement results in massive loading being transmitted to props on the lower floors.

Proprietary props and quick strip arrangements allow the standing support to be maintained while the soffit sheathing can be rapidly struck and re-used.

It is possible to arrange for support of the perimeter of the slab formwork from the beam side members. When such junctions are detailed it must be remembered that beam side members are usually struck out prior to the removal of the slab soffit formwork. Suitable arrangements must be made to facilitate this operation. Where the steelwork in reinforced concrete beams is of a particularly complex nature, the steelfixing is carried out in position, on the soffit, and the side forms erected. Where the programme is tight it can generally be arranged that the formwork gang can form the soffits of all beams and insert alternate beam sides and sheathing supports. The gang can then follow up the steelfixers by inserting the remaining beam sides and filling in the intermediate sheathing bays. Changes in the slab level mainly occur at beam positions and close co-operation is required between the concreting gang and the carpenters at these points. The upstand formers are erected once the concrete has hardened sufficiently to allow working from boards in the placing of such components.

Service holes in beams can be formed in a similar way as that adopted for walling, although the placing of the concrete is more difficult. It is therefore advisable to form holes through the form sides in the appropriate position and to place boards between the opening cheeks when the concreting reaches the appropriate level.

Where beams have splayed sides, it is effective to form rigid frames to support the side sheathing. These frames should consist of diaphragms of ply sandwiched between twin bearers laid on runners. The frames should be sufficiently robust to support the concrete without transference of the loading to the adjacent slab forms. The panels can be manufactured in suitable lengths with pockets formed at the joints to allow the insertion of splayed blocks to the head of the standing props. When striking the panels are dropped from between the props in complete units and, because of the splayed sides, should strike away in sections quite freely.

Cambered soffits can be formed by laying runners to a level line and then planting packers on to suitably spaced bearers to spring the sheathing to the correct profile. This method is satisfactory for the flat curves, often demanded by modern concrete work. Where marked profiles are to be formed, a technique similar to the traditional centering method using sawn ribs with transverse lagging members to form the shaped face proves very effective. Where the finish to the concrete surface is critical, the lagging timbers may be faced with a sheathing of thin ply.

Upstand beams can be formed in the same way as low walls, a kicker being laid and the form sides clamped back to it by proprietary clamps or by soldier members bolted immediately above the level of the concrete. The soldiers should have their projecting ends spaced by timber blocks to provide a clamping action, care being taken to ensure that the soldier members are substantial enough to provide the required clamping action as well as sustaining the concrete pressure. It is also essential that the upstands are cast and matured before the slab soffit supports are removed.

The sequence of constructional operations often calls for the formation of beams cast between wall or main beam faces. It is useful to provide, particularly in the case of short intermediate beams, a striking fillet at one end of the panel used in this type of operation to allow for the removal of the forms after concreting.

This applies equally at all soffit junctions between the secondary beams and the main beams which have the greater depths.

Beam formwork is generally a primary operation in the formation of flooring and should always be treated accordingly, consideration being given to the provision of fixings for the support to soffit panels. There are various ways of linking the operations of beams and floor slab formation. One point to be remembered at all times, however, is that apart from the structural considerations the line and level of the beam soffits is immediately apparent on entering a concrete structure and has a marked effect on the appearance of the concrete frame. This, coupled with the performance requirements, calls for considerable attention during the erection stages with regard to the accuracy of the set up.

11.5 REINFORCED CONCRETE SLAB FORMWORK

Soffit formwork for reinforced concrete floors is generally related to beam arrangements and since many trades can be concerned with the casting of floor slabs, the speed of soffit erection will to a large extent govern the speed of the whole of the concreting schedule. At present it is usual to erect a grillage of adjustable props, surmounted by timber bearers and joists or proprietary beams to support the sheathing arrangements. Sheathing may consist of sheet steel panels or ply panels, either a framed proprietary pattern, or simply unbacked sheets laid over the bearers. Some contractors have experimented by using ply panels framed with $100 \text{ mm} \times 50 \text{ mm}$ or similar on edge to be used both as wall panels and floor sheathing. Unfortunately, while large areas are simply formed by individual framed panels, the weight of such panels does not bring about speedy erection. While the striking operation is being carried out, these panels can become difficult to handle and remove from below the freshly cast concrete.

Proprietary ply-faced steel and alloy framed panels based on preferred modules have proved to be extremely successful and, as with other proprietary systems, the components may be purchased or hired for a contract. A brief description of the methods used will suffice as manufacturers' leaflets and brochures fully illustrate the process and components available.

Beginning adjacent to a wall, or pair of columns from which temporary support can be obtained, adjustable props are set in position and spaced by means of a gauge to the correct centres. These props support beams from which studs register in the plate head of the prop. Sheathing panels are then placed on to the bearer beams and again located over the projecting studs thus forming a sheathing arrangement of close jointed panels closely retained in contact at all edges and ensuring good grout-tight surfaces. Certain systems omit the intermediate operation of forming the bearer beams, and simply place the props at the four corners of the panel, each prop supporting the corners of four adjacent panels. Many systems now incorporate 'quickstrip' arrangements for standing supports, braces and tie members to accurately position the standard. The formation of infill areas around column heads and in areas not suited to sheathing with the standard modular panel is carried out by the insertion of channel- or box-section components which support ply laid in the normal way. To ensure that the striking operations do not cause damage to the metal framing of the form panels, and that the form faces do not become damaged by falls during striking, an easing arrangement of keyed supports enables the panels to be stripped individually from the concrete face by operatives who work from suitable mobile platforms. Each panel is removed by the operation of the key release which allows the panels to be lowered for release, cleaning and re-use. These arrangements are obviously called into use where large areas of sheathing occur on fairly standardised buildings and buildings of modular design. Because of the complexity of the flooring bays in many buildings brought about by specific applications to or through the geography of the site, it is often necessary to revert to the flexible arrangements which can be provided by using telescopic props, bearers, joists and small sheathing units or ply sheets cut to profile as required.

Where beams are cast prior to the slab formation, or where casings are being formed around structural steelwork, it is often possible to erect formwork without any intermediate support from the previous floor slab. There are several systems of proprietary adjustable centres available which are capable of providing large span supports in this way. They can be arranged to take their support from plates or brackets attached to previously cast concrete or precast beams, the beams in turn being supported from the beam below. The centres are used to support transverse joists under the selected sheathing material. Lattice centres are a natural development of the shorter span telescopic centres frequently used in small bay formwork, either supported by nibs on to the previously cast beam from runners supported by bolting through the beam, or by bracket hangers suspended from brickwork at the perimeter of the slab. Contractors are particularly attracted by the use of centres in that the working area below the deck can be cast free of obstruction from such items as props and braces. Centres are simple to place, since they can be dropped in groups by crane and subsequently skidded into place as

FIG. 11.5. Trough and waffle floors have much to offer to both the architect and engineer. The soffit of this floor is typical of modern practice with the use of moulded formers.

detailed. However the main difficulty with centres is their removal from below freshly placed decks, and often some ingenuity is needed to effect their removal.

Where concrete bays are open ended and not curtailed by an edge beam, the process can be aided by extended runners taken outside the structure. Centres can be dropped a few millimetres by means of wedges or pages, slid along these runners and then bolted to the beam until they are in such a position that they can be lifted by crane into the next position.

An urgent safety consideration requires that props must not be inserted under centres, particularly those of lattice construction since the redistribution of loading may cause the centres to fail. Centres must always be used in a simply supported role.

A camber can be produced by particular designs of centres and it is necessary for the person who details the layout to ensure that any eccentricity of camber, due to adjustment, is catered for in the bay layout.

Formwork is always formed in continuous runs as this allows the greatest area to be sheathed without any necessary modification to the sheathing size. The perimeter can be filled in by lap panels or cut ply sheet to form make-ups. Bearers can be kept in the longest possible lengths without the need to trim by forming lap joints over the tubular supports and staggering the panel arrangement to suit. Where proprietary steel systems are used, without a quickstrip facility, standing props may be inserted either immediately under panels at the specified centres, or under the strip make-ups arranged in the required position in the bay. When drop panels are formed around column heads, the supporting arrangement of props, bearers and joists can be erected under the drop sheathing areas and fully

Fig. 11.6. Extreme care is needed to support wood wool which is used as permanent formwork.

sheathed. The main areas of sheeting can then be laid in the normal way, between the platforms so formed, and make-up panels can be fitted between the perimeters of these areas. Upstand fillets, formed of rebated timber or timbers spiked together, can be used to provide a land of suitable depth to accommodate the thickness of sheathing material.

For some work it is necessary to provide grillages which support precast or prestressed flooring units with hollow tile or concrete intermediates and *in situ* concrete margins. This may be readily carried out by running rows of props and bearers in the appropriate direction, the bearers having a rip or packer of thickness equal to that of the sheeting material selected to form the soffits for the *in situ* margin. When such margins or intermediate beams need to be sheathed, the packing strip can be removed locally and the sheathing laid over both the supporting grillage runners and the extra intermediate joists as dictated by the slab thickness. Where solid concrete floors are cast with profiled soffits or ribs which form beams, the sheathing laid over the runners is often composed of boards of suitable width to form the soffit of the rib or beam. At the same time these boards can support form panels of pressed steel or reinforced card, the former being sprung from the concrete on striking and the latter torn out.

Trough and waffle formers now constitute a major area of form technology. Engineers have welcomed the economies offered in the medium span range and architects have been quick to realise the advantages of coffered ceilings for effect and utility. Preferred sizes have been established and it is possible to hire or purchase glass reinforced plastics or moulded plastics formers in a variety of sizes. There is always a demand for non-standard components and when these are being prepared considerable time in the sorting and identification of sizes can be achieved by the inclusion of a colorant in the plastics moulding.

Ideally the support system should be co-ordinated with the coffer arrangement and this is essential where any quickstrip arrangement is to be incorporated.

11.6 STOPENDS

Stopends and the construction of day joints in concrete are items which if not correctly planned and executed can become quite costly. At present a considerable use is made of steel and plastics water bars at joint positions. In many cases this demands a joint in the stopend material to accommodate the projecting portion of the water bar. This joint, together with the joints which accommodate

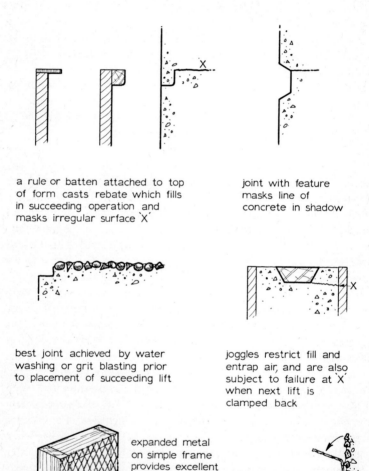

a rule or batten attached to top of form casts rebate which fills in succeeding operation and masks irregular surface X

joint with feature masks line of concrete in shadow

best joint achieved by water washing or grit blasting prior to placement of succeeding lift

joggles restrict fill and entrap air, and are also subject to failure at X when next lift is clamped back

expanded metal on simple frame provides excellent surface for jointing purposes

expanded metal peeled away from face

FIG. 11.7. Stopends and day joints.

projecting steel, often results in a four-piece stopend. It is advisable to screw the individual components of the stopend from the back of the bearers, or at least to nail from the back, to enable the fixings to be withdrawn prior to striking out the stopend material. The more traditional method of breaking out stopends after casting can otherwise prove to be expensive throughout a number of uses. With material correctly cut around the reinforcing steel, there should be little or no grout loss, resulting in honeycombed concrete, and the stopend will act as a jig in the steel positioning operations.

Joggle joints, if specified, can be formed by using sawn timber, or machined to profile, and where large joggles are required, bolts can be arranged to jack out the joggle formers after the concrete has been cast.

With the engineer's approval, inflatable rubber or a plastics duct former provides an effective joggle former if inflated between the two layers of reinforcement which project from the walling. Suitable checked fillets maintain the correct cover between the form face and steel. Within complicated reinforcement cages, it is better to use natural timber rather than plywood. Plywood can prove to be difficult to split out because of its laminated construction. Alternatively hardboard can be used as a facing for immediate contact with the concrete, since it can be left in the face after the stopend carcass has been removed and struck during a subsequent operation.

With respect to large casts, and after discussion with the structural engineer, expanded metal on an open framework provides a useful stopend or dayjoint former. Little or no fines from the mix percolates through the mesh and when it is torn from the concrete face there is an excellent keyed face which, after washing and brushing, can provide a mechanical connection between the two concrete bays.

As a general rule, where the specification permits, the sooner the stopend or dayjoint material is removed, the better the final results. Early removal can be achieved even when deep lifts have been cast since this allows early preparation of the face for the connection work and subsequent operations. Bush hammering and grout coating techniques are gradually disappearing and an enlightened approach to joint formation is becoming much more evident. This is certainly the case for building work, as 40% of the formwork labour is in the formation of stopends and dayjoints to lifts and bays.

Where continuity steel and projecting connections are regularly positioned, and repetitive use of stopend materials is possible, these can be framed and thus readily attached to the adjacent form panels. Dog toothing of ply-end formers allow two-part stopends to be extracted from around the steel.

The fixity of stopends is critical and is an important feature in formwork designs. Calculations are often carried out for the sheeting thickness, spacing support and similar items while the stopend arrangement, important as it may be, is left to the individual operatives to construct virtually in position.

Sound cleats which are bolted to the form face must be provided, and the backing members should be designed to offer a clamping arrangement at the position of the stopend. In slab-forming there should be an adequate support to avoid any sheathing deflection which allows grout to pass between the underside of the joint former and the sheathing face.

On exposed work all stopends must be fixed plumb and tight to the form face. The face finish can be improved if rule battens are attached adjacent to the stopend to form a small rebate in the concrete face which, when filled by the fines from the next casting, provides a perfectly straight joint on the concrete face.

Chapter 12

CONSTRUCTION OF MOULDS FOR PRECAST CONCRETE

12.1 GENERAL CONSIDERATIONS

Constructional details for precast and prestressed work are basically similar to those adopted when forming *in situ* concrete work, with perhaps additional refinements to the design and construction brought about by the tighter tolerances involved. Moulds are mainly used under factory conditions with less tendency for damage due to weather exposure or by rough handling between uses. Precast and prestressed concreting conditions demand heavier vibration of the concrete and mould faces have to resist greater scour than perhaps when general formwork is used on site. Mouldwork is usually the product of a works mouldshop or a special sub-contractor's shop. Invariably a higher re-use value than with *in situ* forms is required.

The manufacturer often invests extra money in the provision of mouldwork for fairly repetitive items with the knowledge that properly stored, after completion of a particular unit, he may well be able to modify the mouldwork to provide further castings on work of a similar nature at a later date. This applies in the case of standard products and system components.

If the methods of construction and materials used are maintained to set the standards of finished sizes and grades of material, component parts manufactured into moulds can be so constructed as to lend themselves for re-use with the minimum amount of remaking or modification. Plain side panels to beam moulds, column moulds and piles should vary little in detail and can, if carefully used, form useful items of plant for general use. Mould costs can thus be reduced to a minimum by spreading the prime cost over the various contracts cast using the moulds. While precast and prestressed moulds can be constructed to be as equally robust as forms for *in situ* work, it is necessary to employ a number of refinements and constructional details which are suitable for use where rapid turn round of the mould is required to suit factory production rates.

With the ease of prefabrication of reinforcing cages offered by even the largest of precast units, there are few occasions when after the casting operations have begun, the side panels and stopends are not used in everyday casting. It is indeed essential to the financial success of most contracts that casting is carried out on a daily cycle, so that to this end, manufacturers may double or treble the ratio of pallets to mould side panels and stopends.

This of course depends on the specification which governs the removal of the casting from the base. In prestressed work it also depends on whether strength has to be obtained in order to allow stressing of the unit prior to removal from the casting bed. Where prestressed units are concerned, particularly where long-line beds are used, complete runs of baseboard and side forms are generally laid between abutments. Alternatively a complete base can be laid and a set of mould sides moved down the bed daily as the units are cast. Obviously where there are sufficient castings of similar detail to warrant the expenditure involved, the complete bed or several beds furnished with side forms provides the most economic working which allows continuity of operations throughout the working shift.

With precast and prestressed operations there are seldom more than four groups of skills concerned: the carpenter and mouldmaker, the steelfixer and the concretor and dresser. The dovetailing of the various operations presents less difficulty than than that encountered with *in situ* work. Alternative work can readily be provided when a particular operation is in progress and would, in the case of *in situ* concrete work, tend to delay the subsequent trades. With casting bed space carrying heavy overheads careful planning to allow rapid modification of type of unit is demanded thus calling for close attention in the construction of the moulds.

12.2 PRECAST PILES

Reinforced concrete piles were among the earliest precast products to be produced and with the introduction of prestressing into pile manufacture these items formed a large part of the production of many factories. Precast piles were frequently manufactured in special yards which were set up on major contracts before the last war and, in fact, this still applies with the more inaccessible sites or where large quantities are required. Today, however, thousands of piles are cast in factories and transported by road, rail and barge to the construction site. Methods generally have not altered over the years during which piles have been manufactured, although modern developments in stressing and the casting of hollow units have altered the approach of some manufacturers. Piles are usually cast in gang moulds to reduce the amount of side support required to the minimum; the quantity cast in gang depends on the capacity of the batching plant and bulk of concrete which can be handled in the course of a shift. With many factories the size of gang is limited to the bed space that can be set aside from other operations though of course concreting capacity will be the main factor. In precast work it is necessary to provide a level bed on which baseboards or concrete sleeper soffits can be laid. The pallets, as they are called, are generally constructed of timber of 50 or 75 mm thicknesses securely bolted to the casting deck or of up to 150 mm of concrete sleeper occasionally ragbolted or similarly tied to the casting bed in order to prevent scatter during subsequent lifting operations.

The internal side panels may be formed of 6 or 8 mm sheet steel with suitable lifting eyes welded on to the tops of the plates. The outer side panels can be suitably framed steel sheet strutted and tied in the normal manner. Some casting shops have quick action clamps which bear on soldier members incorporated as plant items. Where casting is carried out on site the struts can be taken back to bolted plates. Site precasting is frequently executed in sunken moulds to allow ease of strutting to the outer sides. Spacers are inserted which accurately position the intermediate forms after the insertion of the steel cages. As the concreting proceeds, the mix being evenly distributed throughout, these spacers can be knocked out to leave a clear area which allows the trowelled finish to be formed. Chamfers can be formed by suitable fillets positioned on the base, and by trowelling or by inset fillets at the head of the mould panel. Lifting holes can be formed by dowel bars passed through the side forms. Where required, vertical holes can be formed by bars passed into the pallet and centred off by battens clamped to suitable fixings on the form sides.

Where delays are incurred while the transfer strength is achieved, the manufacturer often resorts to double banking of piles. This is effected by partially drawing the inner form sides, raising the outer sides and restrutting, laying a paper or plastic sheet on the face of the previously cast layer of piles and proceeding with steelfixing and concreting as before. While this operation can successfully reduce by several days the time during which a bed is occupied by a given quantity of concrete, it should be remembered that the outloading period can be doubled. A more productive approach however is the installation of one of several means of accelerating the curing process which can include steam, electricity, the use of hot water in the concrete, and the provision of adequate heat insulation both below and around the moulds.

The manufacture of concrete moulds has been covered by the author in some depth elsewhere and the reader is referred to the bibliography at the end of this book. It is essential that moulds for repetitious factory or for use on site should be constructed in steel. The panels can be in constant use especially where the manufacture of standard units is a feature of the production. The panels are subjected to heavy vibrational loading and considerable scouring action. Where removal is carried out by crane, the use of a spreader bar is essential to ensure that each mould panel is withdrawn directly from the concrete. Two operatives are required to attend to this operation to ensure that the panels do not bind or become buckled in the process.

Gang-casting operations are often used in the production of piles in works or on site. Where there are considerable numbers of piles, either precast or prestressed, it is almost standard practice among civil engineering contractors to gang cast—generally on the long-line system. Master units are produced and these are used to generate gang-mould units which accommodate four or six piles in their various widths, the units being installed side by side and end to end to comprise the mould. Gravity abutments are cast at each end of the bed so formed. The moulds allow some minimum lead and the piles are initially free from the mould by jacking, being subsequently handled by derrick or gantry crane. Gang casting provides a simple arrangement for mass production because of the ease of concrete placing and compaction. The detailed arrangements must of course vary in each cast but it is evident that the process can be engineered such that production can be carried out using only simply skilled operatives. Transporting costs can be reduced to a minimum by the careful siting of the casting yard and stockyard.

12.3 FLOOR UNITS AND DECKING PANELS

Floor units and decking panel moulds often receive a similar treatment to those used in pile casting and depending on their width and cubic content, are often cast in gangs. In these instances rebates for seating of tiles or concrete blocks are to be formed by using timber—preferably hardwood fillet or steel sheathed softwood fillets attached by bolting or riveting to the steel side member. Wherever possible, a lead should be provided on the feature to assist in the withdrawal of the side forms. Light alloy can often be economically used for such applications and allows of recovery of some percentage of the original purchase price in re-sale as scrap.

Where heavily cambered soffits are demanded, these can be formed by springing timber sheathing over carefully spaced blocks or by the provision of a concrete base formed by erection of shaped mould sides and screeding. This process followed by trowelling, dressing and, if required, final grinding presents a suitable surface for casting. Such bases must be cured slowly and be oiled in the normal way, as for timber or steel sheathing.

Numerous proprietary arrangements are available for forming hollow cores in precast deck and floor units. Among the most efficient is the system where rolled steel sheet of suitable profile is rigidly locked by the insertion of plastics or hard rubber. The fillet is removed from the joint after concreting to allow the metal profile to flex and be withdrawn from within the unit. Formers can be prevented from floating by the use of finger plates in the form of stretchers which span the mould and which can be removed as the concreting and vibrating operations proceed along the bed. Inflatable core formers and soft rubber formers offer other excellent methods of forming hollow cores, although the latter requires support from the steel and inserts to prevent floating or deflection between supports. Further information on formers can be found in the chapter which deals with materials.

Purlins which can be used in conjunction with precast frames are items that can be profitably cast in gang, the actual operation of placing concrete being made easier by the increased area over which the concrete flows prior to vibration and finishing. Individual purlins present little cubic content and, moulded individually, may well consume as much time in placing, in order to avoid spillage, as the

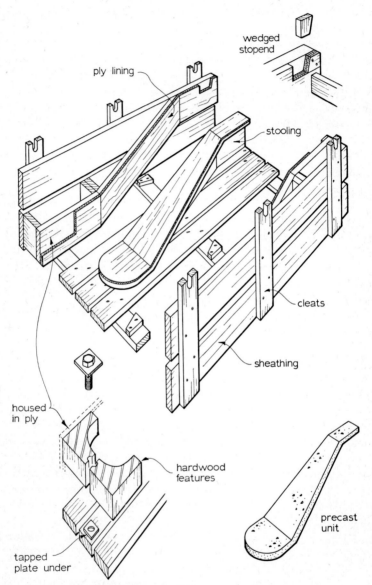

Fig. 12.1. Typical precast product mould in timber and ply.

far larger units. Where the section is regular and is not unduly complicated by end variations, such units can be readily formed in a manner similar to that adopted for concrete piles and floor units. Numerous small products such as cattle slats and piggery floor units, silo staves and such like can usefully be cast in gang moulds. Where the size of component allows the ganged moulds can be used in a mobile capacity, the moulds being circulated through making, curing and de-moulding stations where necessary, and turned for the de-moulding process.

12.4 MOULDS FOR MASS PRODUCTION

There are of course specialist firms whose entire works output is composed of precast units for specific applications such as patent flooring or structural elements. The moulds which are used in such factories are built of steel, robustly fabricated to provide constant accuracy of profile and multiple use. Carefully designed mixes allow a daily casting cycle. The moulds are designed to be operated with the minimum expenditure of labour in striking and re-erection and have side forms hinged to the base. Built-in clamp or tie arrangements and pneumatic or hydraulic systems are used for securing the forms during concreting. The moulds for such applications are generally the product of a specialist mould contractor and it is useful to identify some of the critical factors that govern the installation and operation of such equipment. The unit mould used in the manufacture of proprietary or system components provides a useful instance within which the principles of the more sophisticated steel mould may be discussed.

Trends within even the more traditional precasting establishments are now towards mechanisation, and the mould and its ancilliaries offers considerable opportunities for reductions in labour employed.

In order that mechanisation can be adopted, a discipline must be imposed for the production of a range of standard components. With standards of width, length and thickness established, quantities of mould accessories, sides formers and the like can be reduced. Fewer variety of accessories means that more parts can be attached to the basic mould bed and thus there is less handling of units at the time of de-moulding and re-assembly.

Simple mould components, in many instances fixed to the basic bed, suffer less damage and are simpler to maintain so that a good standard of accuracy can be maintained. The moulds can be heated in a variety of ways and thus the output from a given quantity of mould equipment can be increased. Simple mould beds lend them-

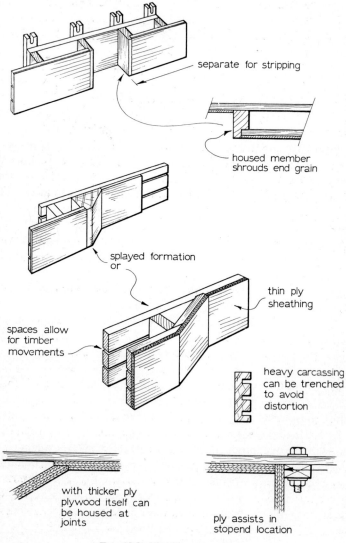

FIG. 12.2. Mould construction.

selves to rapid changes in basic production—a factor which can be critical in the satisfaction of demand within the precast frame industry.

FIG. 12.3. Typical precast units in workshop construction. Simplicity of line allowed the adoption of concrete mould pallets and permitted multiple re-use of mould sides by modification between uses. All corbels and projections were accommodated in basic mould panels.

12.5 MOULDS FOR FRAME COMPONENTS

With present day trends towards precast modular building, a large volume of the product of the average precast works can be made up of simple column and beam formations. Experience gained in erection methods and developments in available mobile plant have resulted in large capacity cranes being readily available in most parts of the country, so that the size of the component parts has increased. Columns are often cast to extend through three and four storeys of a building, with suitable nibs and projections to form seatings for intermediate beams and floors. There is a tendency now towards the production of composite components, i.e. H frames and multiple T floor units. Large-span beams, including segmental components, can be precast in works and transported by way of specially constructed transport to site. With the heavier sections and larger units, a great deal of precast work is taking on a similar aspect to single lift castings such as is adopted for *in situ* work.

When large precast units are manufactured special measures with regard to the maintenance of accuracy are necessary. The provision of manufactured concrete pallets, or concrete lined with timber, overcomes several difficulties in the formation of long precast units. Where more than one crane is used for lifting there is always a tendency for the base material to scatter which causes delays through relaying and lining-in the bases. When the pallet material is timber, difficulties can arise in maintaining the line of the base when side forms are wedged, cramped or tied across the base. Timber is bound to absorb moisture from the concrete mix or lose moisture through drying between uses, with the possibility that winding and twisting setting will permanently upset the accuracy of the base. Lined or unlined concrete sleeper walls, or even walls sheathed with steel for multiple re-use, act against these tendencies and ensure good accuracy of the finished product. When concrete bases are being laid no adjustment need be made to the through-tying method, if this is a standard procedure. Roughly formed through-holes or holes ferruled within a conduit provide a suitable route for the lower ties in the system.

12.6 MOULD CONSTRUCTION—GENERAL

As described earlier, moulds for precast work can be constructed to the optimum profile, care being taken in the selection of the face for contact with the pallet so that deep pockets do not have to be formed for projecting nibs. This consideration of 'aspect' or 'way up' of casting is critical for the success of the whole operation. Sheathing depends on the specification which governs the finished product. Many manufacturers find that timber form panels made on standard lines offer the greatest economies when lined with thin plywood sheathing which can be replaced, as required, to provide the maximum number of re-use operations from the framed backing members. Ply facing allows the backing members to be spaced slightly apart and thus eliminates buckling or bowing of the sheathing due to swelling and shrinking as the backing material loses or gains in moisture. Heavier ply sheathing on framed backings of 100 mm × 50 mm, or similar at larger centres, is favoured by many manufacturers since the ply can be fixed by screwing to facilitate turning when the face becomes unduly scarred or scoured through the placing of concrete. For small-section units, single vertical cleats can be used for tying, though as depths increase these are replaced by twin soldier members which are spaced by screwed or nailed blocks. As with the majority of formwork these members can be of standard section and will generally be understressed, their spacing on the mould being decided by the casting bed, grillage support or profile of the unit being cast. Again, while smaller sections generally could be used, 100 mm × 50 mm and 100 mm × 75 mm bearers and the 40 mm and 50 mm nominal sheathing thicknesses can be incorporated in the manufacture of the panels. If more accurately designed timber sections were used the methods of handling and striking would probably cause wracking or wind. Similarly where nibs and projections could be housed within a profiled panel which consists of one depth of sheathing, they are generally housed within recesses formed by a secondary layer of sheathing or partial sheathing, which serves to reinforce the mould panel and make it capable of being struck from the face without damage. While offering a further example of 'over design' these extra layers of sheathing offer excellent opportunities for the mould panels to be made adjustable in length and for the provision of stripping fillets. The secondary or partial sheathing should be carried past projections or returns to the distance of one or more bearer spaces.

Tie-bolt arrangements vary between firms and are decided entirely by the designer bearing in mind the method most suited to the existing working arrangements on the casting bed. Some factories have built-in anchorages for plates while others have grillages consisting of RSJ or channel section steel from which vertical soldiers or clamp posts can be obtained to help with strutting back side moulds. When a casting bed is laid it is useful to form a complete grillage of buried anchors, ensuring that the holes are filled with timber plugs until required. Where clamp posts are installed with screwjacks, for plumbing and lining sides, these should be fitted with cramps which have square-cut threads to avoid clogging and thread damage by grout or concrete spillage during casting.

There are several key points to be noted, regardless of the actual arrangement of tie bolts or cramps, with respect to the tying arrangements. Whenever possible, to facilitate trowelling, there should be a clear space of 100 mm minimum between the top of concrete, as cast, and the underside of the tie. End ties must be arranged to coincide with the stopend position to prevent slight deflections in the sides

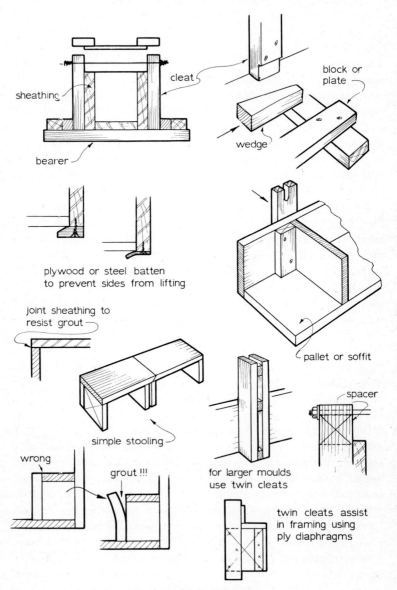

Fig. 12.4. Mould construction.

which might result in the stopend and side panel parting. Such a parting movement would allow grout to escape and the build up of pressure at the joint during placing would cause further deflection and subsequent distortion of the unit. Equally important, is the adequate tying back of panels which pass or overlap where there are variations in section. Ideally, features and fillets should be recessed into the side forms to present butt joints at the concrete face, and this can be readily accomplished where a thin ply sheathing is used over timber panels. For I-section work, this method of construction allows the feature to be moved along the face or replaced during any modification of the mould assembly. The intermediate panels of sheathing can be removed and adjusted to provide recesses suitable for the modified feature blocks.

While on the subject of *in situ* column forms, the question of butt-jointed corner fillets which are integral with the form panel has already been discussed. The same considerations govern the features and fillets of precast moulds. It is quite simple to incorporate solid chamfer fillets with the mould material, or even to let them into the sheathing material. It is slightly more difficult to trowel off, as small upstands are formed to the top edge of the precast moulds, but the increase in uses of the fillet, and the improved quality and accuracy of the chamfer on the product, amply repay the slight amount of extra labour involved.

Many manufacturers use through-tie rods at the bottom of a mould assembly, and accommodate these by housing grooves in the sheathing, bearers and cleats to the pallet. An extremely efficient method of tying is to use blocks and wedges, or plates and wedges, to retain the bottom of the side panels which are in contact with the pallet sides. Where this method is used the bottoms of the cleats are splayed out to suit the wedge angle, preferably at a slow pitch to avoid any jumps during placing. Blocks are screwed, or screwed and glued, to the baseboard cleats in the appropriate position to allow some pinch when the wedges are halfway engaged. Where concrete sleeper bases are employed, a plate secured to the casting deck replaces the blocks used in this system. Small blocks or steel cleats fixed to the bottom of the side sheathing engage under the soffit and prevent the side member from climbing while the concrete is being placed and vibrated.

It is often necessary to combine the tying operation with the holding down of the side mould, and a useful, quick action clamp which can provide both facilities can be manufactured from steel plate. The plate, which is slightly set and which has a clearance hole to suit standard soffit bearers, has a tapped hole in the upper section to allow the insertion of a threaded bolt with a spigot end. When in use the cranked plate is slipped over the pallet cleat and the bolt tightened to clamp back the side bearer to the edge of the pallet. The spigot projects through the cleat and into a suitably placed hole in the edge of the pallet and thus prevents any tendency to lift.

Spacers are usually required at the head of a precast mould and should be accurately cut and provided with a small checkout which rests on the head of the side cleat and thus prevents movement during concreting. Alternatively, lengths of barrel slipped over the tie rods, with washers to spread the end-bearing pressure uniformly around the slot, or drilling at the head of the cleat present a positive spacer which can be quickly inserted.

Where the concrete units to be cast present a number of faces in various planes, it is essential that construction begins at an overall grillage level formed by the basic members of the panel. Directly attached to these is the sheathing to the outermost face, for example the face of the flange of an I-section unit, or the face of a nib or corbel. Blocks are added after this to locate the next plane of sheathing in its correct position, and this is continued until the main faces are formed. Inserts can be placed to form the splayed faces where these are wider than can be formed by the shaped edges of the various sheathing members. Blocking pieces or diaphragms can, where possible, be sawn from single timber pieces. Where features are particularly deep, ply diaphragms or framed timber diaphragms can be slipped between twin soldiers, or skew-nailed, or blocked on to solid bearer faces. The features should be formed in such a way that the edges of the facing material are masked by the sheathing of the cheeks of the feature. The smooth face thus formed avoids any difficulty in the striking operation brought about by infiltrating fines, or binding on edge grain. Large features may be bolted back to the main panel assembly in such a way that the bolts can be removed when the side forms are struck. The feature formers within the aperture may be withdrawn by re-insertion and a jacking action of the bolts.

Where a number of nibs and projections exist on the faces of units, and vary in position on various types, the mould panel can be manufactured to the overall profile which contains the variations.

Construction of Moulds for Precast Concrete

Infill pieces or stools can be inserted to complete the profile for particular units, and these need to be split to allow withdrawal from between the return faces which would otherwise tend to trap the stools and make them difficult to strike from the concrete face. Through-holes in the units can be formed by the use of dowel pins of the appropriate size, passed through the faces of the mould. The side panels must be reinforced at such points to eliminate wear and consequent misplacement of the through dowels; small pins can be used in pairs to locate large formers in order to avoid massive holes in the mould face.

Where there are large service openings through concrete beams or columns, they should, wherever possible, be formed on the vertical axis of the unit as cast. In this way it is easier to ensure that there is maximum compaction since the vibrators can be passed into the concrete between the core member and mould face in the normal way. There is also less tendency for air pockets to form which will need to be patched, or for repair work to be carried out on the concrete unit when the unit is struck. However, it is necessary for striking purposes to provide an adequate facility in the form of stripping fillets or folding wedges which provide ample clearance for core withdrawal. The core-box sides should be allowed to pass into the thickness of the baseboard to ensure a sound butt joint at the junction of the core and baseboard, and to exclude any possible leakage of grout.

Where core holes pass laterally or longitudinally through the mould and unit, and it is impossible to re-orientate the former by turning the unit, special arrangements must be made for filling and compacting. Large formers can be arranged so that the concrete can be placed via an opening in the mould side. Where an opening cannot be introduced, a careful sequence of casting must be laid down to ensure that the concrete is fully compacted under the former prior to the placing of concrete above the soffit level, since this avoids air being trapped.

12.7 STOPEND CONSTRUCTION

Stopends in precast concrete work require particular attention as they can be re-used repeatedly in varying positions within the moulds. Numerous variations of stopend fixings are possible, but it is necessary to use a method which while providing positive location of the stopend also ensures complete freedom of movement within the mould to cope with length variations.

Stopends should be carefully designed as in very many cases they are used to form connecting faces, or to locate projecting reinforcement or bearing plates which govern the accuracy of the finished

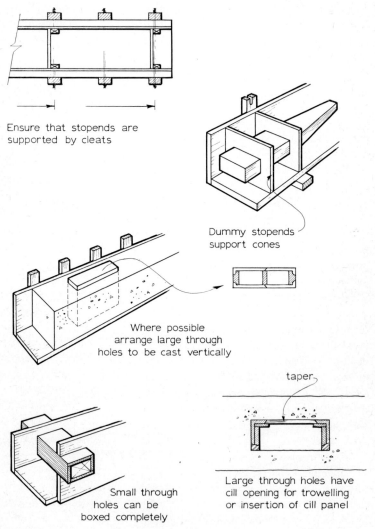

Fig. 12.5. Mould construction.

structure. The slenderness of the stopends and their vulnerable nature quite often dictates the selection of steel in their manufacture. The use of steel stopends allows welding or bolting of cleats and jigs, and the location of projecting features and the stability of the material permits a designer to use the stopend as a diaphragm which governs the section of the mould and thus maintains accuracy.

Where there is considerable amount of steel projecting from the end of a unit, or where it is critical that reinforcing bars should be positioned accurately, it may prove economical to provide disposable ply stopends on a steel frame stopend. In this way the ply acts as a template, and serves as a protection to the green concrete and is wrecked from the unit prior to erection.

All stopends should be so fabricated as to eliminate the possibility of being wrongly assembled in the mould thus causing inaccuracy in the units. This applies particularly where the handling of units is achieved by stopend profile.

12.8 BRIDGE BEAM MOULDS

Precast and prestressed concrete bridge beams present complicated profiles for the mouldmaker to achieve, since they often consist of the interrupted I-section, unequal I-section or splayed I with anchorage block positions and buttresses, so formed to permit through passage of transverse stressing cables. When such units are used it is desirable to construct a set of basic mould sides which consist of bearers that link a substantial framing, clad with sheathing, to form those faces which are continuous along the length of the unit. Where necessary, diaphragms can be used to maintain these faces in the correct relationship to one another. Suitably profiled blocking pieces are assembled on these basic panels to form the face of the buttresses. Infill panels can be inserted between the frames to provide the appropriate feature profile. By doing this it is possible to adjust both the feature length and buttress position. This method of construction is particularly useful in forming variations in the unit profile such as are brought about where skew bridges present varying pitches of through ducts and anchorages. With such members there is often a cambered profile to be formed to the base of the beam and it is very economical to provide a plane mould of spaced timber sheathed with thin plywood.

These moulds are brought to profile by end bolting to the casting deck or grillage over a series of accurately machined shaped blocks, set at regular centres along the run of the unit.

Where an extreme camber is required, it generally proves more practicable to form bases and pallets of two layers of sheathing, the basic sheathing being open jointed and having blocks screwed or nailed on it or, where necessary, framed up diaphragms to provide the required profile when the second sheathing layer is formed over them. With all such bases, and also where deep bases are formed to accommodate projections from the face of the unit within the thickness of the pallet, there should only be contact between the pallet and the side member at two points. The contact at the pallet sheathing position ensures a grout-tight joint while that at the tying or wedging position provides ease of plumbing of the side mould panels. Continuous contact over more than 100 mm of pallet thickness or diaphragm is undesirable due to the tendency for small particles of concrete becoming trapped, thus upsetting the form joint and the set of the side moulds. Concrete sleeper bases can be set back 12 mm behind the pallet sheathing line, and profile blocks to cambered bases should be kept clear of the side face, with only a register fillet on the sub-structure contacting the side mould to assist with spacing and plumbing.

12.9 STACK CASTING

Where concrete beams and columns are rectangular in section, they often can be cast in layer fashion or stack-cast vertically, economy being gained from the reduction of the number of required bases. This applies particularly to portal frame units, A frames, H frames and combined column and beam formations which are used to support electrical transmission lines. The method entails the careful provision of an accurate concrete sleeper pallet to the required profile. This can be cast within a formwork of square members braced to present a rigid mould and weighted, or bolted down, to the casting deck to eliminate movement. The surface can be trowelled carefully, cured, then oiled with a good quality mould oil. A conduit can be laid where tie rods are required to support the toe of the side panels. For this type of work, it is economical to use heavy section sheathing which allows increased spacing of the through-ties. Side moulds and stopends can then be erected, the bottom tie passing through the prepared hole in the concrete base.

After concreting, the upper surface of the unit can be covered with paper or plastics sheeting and the side panels raised to provide another casting immediately above the first; this process can be repeated several times. When necessary, a unit can be removed from the stack and set up elsewhere on the bed for further casting using another set of mould sides. This method depends for its success on a first quality trowelled finish being obtained on each unit cast. Several variations on bolt or tie rod positions are used, but undoubtedly the most positive system is that in which a tie bolt is used at each soldier position through the pallet or previously cast unit, a sleeved bolt or coil tie through the unit being cast at the appropriate level providing a tie for the next operation. Lightly fixed slotted twin soldiers can be used in such a way that a one-tie fixing and cantilever effect provides a mould which is free of obstructions at the top face of the unit and which facilitates trowelling. The accuracy of the unit in such a method depends on the exact thickness of the pad between the bottom of the twin member and the unit, and any slight offset caused by lodged grout or stones can present inaccuracies.

12.10 DUCTS FOR TENDONS

Where cable ducts are to be formed in precast units by the insertion of foil, rubber or card tube or section, there is a constant need for a means of restraining the duct formers from floating with the concrete, or from being deflected during placing and vibrating, and to ensure their correct positioning within the concrete mass. Several means may be adopted to position these ducts and among these are welded grids or sections of mesh, through dowel bars, wire hangers and concrete diaphragms. The engineer's approval will be required of the method adopted for particular work, but the use of precast concrete diaphragms is a method which is equally suited to large and small sections of concrete as is evident from the illustrations in this book.

For the smaller diaphragms required in bridge beams, a diaphragm thickness of 25 mm to 40 mm is sufficient. Flat moulds are constructed with planted rims which have removable sections to facilitate striking of the diaphragms after curing. The duct openings can be formed by hardwood or metal blocks which are set into the base of the mould and which terminate flush with the sides to allow trowelling of the exposed concrete face. Small sections of the diaphragm can be allowed to project through to the face of the mould panel, to the actual unit, to ensure retention of the cables in correct relationship to the unit faces. One vital part of this method is that the diaphragms should be marked with an identification number to ensure that they are correctly placed along the run of the duct within the unit during the threading operations. Wire fingers, dowel bars or simply the pressure of the mould sides locate the diaphragm within the mould.

12.11 PRECAST STAIRS

With the current accent on precasting as an aid to the maintenance of tight concreting schedules on site, manufacturers are often called upon to cast items such as flights of stairs which have previously been the province of the site concretor. Such manufacture may be approached in a variety of ways, but methods which have proved to be entirely successful are the casting of flights of stairs in lengths between landings by the construction of moulds which have the pallet shaped to the stair profile, and which are edge cast. Alternatively the treads and risers can be inserted into what is virtually a standard mould unit with regard to the placing and compacting of concrete, the various combinations or rise and going being accommodated by varying stoolings.

The first of these two methods is generally accepted to be the more economical method with respect to speed of placement and consolidation of concrete, although the latter offers a straightforward method of producing flights which have a special mix for the surface of the treads and risers. Where upstand trimmer beams or kickers for baluster walls are to be formed, the tread and riser stools can be stopped back to the face line of such upstands. The method of casting with the unit cast 'as placed on site', with riser boards and cut strings forming the side panels, presents difficulties in concreting and trowelling although it is a simple matter to alter the rise and going by using this method of formation.

12.12 TEXTURED FINISHES

Textured finishes for precast products demand special treatment of the mouldwork and dictate the faces which are in contact with the pallet during casting. Treatments have been evolved to produce various textured and exposed aggregate faces; plastics and extruded and rolled rubber surfaces can be used as sheathings to the pallet member of the mould to produce rippled, ribbed and textured surfaces. Timber fillets can provide various shapes to suit some particular requirements, also sheet plastics, vacuum formed over patterns, can be used to form some feature which is unobtainable through other methods. Hardwood fillet and extruded light alloy sections can be utilised to form continuous patterned surfaces.

Where exposed aggregate work is required there are a number of methods available some of which can prove more effective than others. Retarding agents can be used on the mould face and the face fines brushed clear after the unit has been struck from within the mould. Absolute care is required to ensure complete regularity of exposure. With larger sizes of exposed aggregate the sand bed method is extremely effective, if somewhat expensive, with regard to labour. Using this method, a flat soffit of steel, timber or concrete is laid and side moulds erected which enclose the required area. Next, a layer of damp sand is spread evenly over the area and stones placed on the sand, the depth of insertion being arranged to present the required amount of face texture. Exposure can be governed by the depth of the sand layer, and where this exceeds 4 mm it is necessary to tamp the sand evenly to ensure a compact mass thus preventing displacement of the stones during the next operation. The laying of the cement/sand matrix can be based on coloured cement or stone dust. The backing mix follows next, the reinforcement cage being placed when the appropriate cover has been laid. Vibration must be carefully gauged to prevent the backing mix from percolating through on to the face and spoiling the finished effect. For best results the side moulds should be arranged to present an open area for the placing of concrete, the insertion of reinforcement and the trowelling operation. Another method of producing exposed aggregate units is to cast in such a way that the aggregate to be exposed can be placed into the unit on completion of the casting process normally applied to precast work, being rolled or tamped into the face. In each of these methods after the panels have been cured and struck they can be brushed down with a wire brush to remove the sand particles and etched as required to clean the exposed faces of the aggregate. Considerable work has been published on decorative concrete faces and the reader is directed to the bibliography for further reading.

12.13 DUCTS, CULVERTS AND SUBWAYS

In view of the frequent delays in groundwork caused by inclement weather, and the fact that the success of many schemes depends on the contractor maintaining his programme at this stage of the work, precast manufacturers are currently called upon to produce sections for subway or duct work. These units would previously have been regarded as site work and carpenters, steelfixers and concretors would have worked, often under the worst conditions, to maintain progress. In the precast yard or precast works, however, such items can readily be cast in sections with projecting steel, for continuity purposes, or ducts which incorporate post-tensioning on site if required.

Where shallow ducts are concerned it can be a straightforward casting operation with suitably shaped mould panels and profiled stopends which support the inner side formers. Depending on the concrete outline, it is usual for the whole unit to be cast monolithic up to 2 m of inside depth of the wall face. Where deeper units are concerned it is advisable to overcome the difficulties that can arise in the tying arrangements by casting the base slab section as the first operation exactly in line with the *in situ* concreting methods, at the same time forming kickers for the wall section complete with approved joint or weather bar. One or both walls can be added during a second operation. On very large units it is straightforward to add these walls singly and thus obtain extra re-use value from the mould panels.

In the case of tunnels and subway units, the roof can be cast as a final operation, the mould fixing being taken from cast-in sockets or similar tie arrangements.

Segments for thrust boring, and indeed for quite large subways cast segmentally, can usefully be cast on end. The end casting arrangement simplifies the filling and compacting operations.

Where, as in flame barrier ducts required in airport construction, the invert slab of the duct is laid to falls, it is advisable, where the fall is constant throughout the duct sections, to form the fall within the pallet of the mould and the construction joint between the slab and walls in the normal horizontal way. The walling panels can then be manufactured with the sheathing joints suitably arranged, so that as the casting proceeds, sections may be removed to allow the casting height to be the actual height of mould face thus facilitating the insertion of fillets to form the rebates which in turn form bearings for duct cover slabs or grilles. Dowel holes through the wall section may be used in the lifting from the casting bed, or failing this method, buried anchors allow the bolting on of a substantial lifting frame.

12.14 PRECAST PRODUCTS

Small precast units and standard products are generally produced from purpose-made steel or glass reinforced plastics moulds. Kerb units, canopy units, fence posts and the like provide areas where the skill of the individual mouldmaker makes itself most evident. Many contracts call for small items of precast work which, whether cast in works or on site, present particular mould construction problems.

Contemporary architecture often calls for shaped coping units, flying balcony slabs with flueing soffits and similar items. The imaginative designer of precast concrete structures is often presented with problems that involve complicated mouldwork for geodetic construction, space frame units and the like.

12.15 TILTING FRAME MANUFACTURE

Mention must be made of the rather specialised equipment used in casting cladding panels where a great degree of repetition is required. Tilting tables are used for this purpose, either to elevate the units for washing, or aggregate exposure, or for various means of removing the casting whilst eliminating bending from the unit.

The tilting table may be simply elevated by mechanical or hydraulic means to an inclination of about 80° for the removal of the cast component. In some circumstances the table is rotated between 90° and 180° to allow transfer of the panel on to a bogey for removal to the stackyard.

Tilting tables allow economies in the quantities of special decorative aggregates which may be laid as thin face veneers. The aspect of casting, flat cast as it is called, allows the incorporation of layers of insulation, either as face treatment or as a sandwich between the layers of concrete.

Tilting tables are a production tool and generally incorporate some means of accelerating the curing of the concrete.

There are no rules which govern the design and construction methods for mouldwork. As in the case of formwork the designer has a multiplicity of materials and techniques at his disposal; he alone can make the final selections and incorporate them into a system. The eventual proof of the selections made by him will be evident from the financial outcome of the work.

Chapter 13

CONSTRUCTION OF FORMWORK FOR SPECIAL APPLICATIONS, ARCHITECTURAL FEATURES AND SCULPTURE

Where a designer uses concrete as a means of expressing form and texture, and where the plastic nature of concrete is being expressed, a variety of materials can be used as sheathing to impart special textures and finishes to the model.

The formwork designer must ensure that he fully understands the requirements of the architect and designer, and it is essential that the details of a particular model or feature are discussed at length. The formwork designer often has to advise on such matters as concrete pressures acting within the container, and the means of achieving particular textures by the use of commercially available materials. A further aspect, so often overlooked, is the means of stripping or striking the form and the avoidance of unintended geometrical complications.

13.1 SCULPTURE AND LOW RELIEF WORK

Usually the designer, or artist, prepares a face or sheathing from glass reinforced plastics, moulded rubber or plastics, vacuum formed plastics, foamed materials or any one of a large number of materials capable of imparting special finishes. The model may indeed be carved from clay or moulded in plaster.

The formwork designer is required to arrange support for this sheathing or form and must be prepared to advise on the means of handling, placing and compacting the concrete within the form.

Ideally, the casing or container should be backed by an inert material, plaster, foam or even concrete to provide a uniform square, rectangle or regular curved surface. The formwork designer can then arrange to support this surface using traditional methods which incorporate soldiers, walings, raking supports and similar components. It is advisable in the majority of cases to ignore the mechanical value of the casing or modelled form, although the mass of the container must of course be considered in any calculations made for supports, bracing or load transferred to other parts of the structure. It is necessary to establish a clear understanding on the curing arrangements and the time of removal of the formwork, bearing in mind that these arrangements may vitally affect the colour and general texture achieved in the model. An important aspect of the work is that of ensuring that the operatives appreciate the need to care for the mould components and treat them in a different way to the normal form component.

Sculptured work is often provided in the form of concrete panels for incorporation in the structural frame. In these cases the formwork designer has to establish the capability of the panel for sustaining load and pressure, and designing a suitable grillage or carcass to support the panel during concreting.

Disposable moulds such as sculptured foam present little problems; the re-usable panel however requires every care in cleaning, oiling and the application of parting agents to retain the pristine surfaces and fine modelling incorporated by the artist.

Fortunately designers who work with concrete make a study of the technology involved and indeed many are experts in this field.

13.2 IN SITU MOULDED SURFACES

Where heavily moulded surfaces are to be cast *in situ*, the formwork designer must pay every attention to the detail of form construction, since the form approaches the precast mould with regard to the degree of complexity. The connection between the component parts of the sheathing, the fixing of the sheathing to the backing and carcass must all be subject to careful scrutiny. Failure to achieve sound construction and fixity while resisting grout infiltration and containing pressures, can result in expensive refurbishment or replacement of expensive parts of the form.

Heavily modelled surfaces may be built from rebated, moulded timber, framed ply panels or of laminated selected timber. The way in which these materials are combined and used in conjunction with the backing will determine the number of uses achieved. Glueing or bonding, and screwed or bolted connections, should be used and each material put to the use for which it is best suited.

For highly moulded or deeply relieved surfaces moulded rubber and polyurethanes provide ideal surface materials, the polyurethanes having the advantage that they may be used in thin layers bonded to inert formers or in substantial sections extended where necessary by filler blocks of foam, the whole cast around an armature of steel which can be used to facilitate fastening to the carcass or backing. The flexible nature of the rubber compounds and polyurethanes is of value where internal vibrators are used for compaction and is useful when castings are produced from fairly heavily constructed forms.

13.3 GENERAL CONSIDERATIONS

The mould oils, parting agents and retarders used in producing special finishes must be selected with care. Generally, the more expensive chemical parting agents provide the most satisfactory results, especially where the application is carefully controlled. The development of surface retarders has reached the stage where these can be regarded as an extremely positive and reliable means of achieving exposed aggregate surfaces. Lacquer-type retarders provide a regular degree of retardation and when used in conjunction with the coating medium provide almost complete protection for the form from scouring, scabbing and general degradation of the form face which are apparent when deeply featured forms are removed from between near parallel cast concrete faces.

In sculptured and featured work, the designer frequently wishes to differentiate between smooth and textured concrete and perhaps to express the colour of different aggregates in adjacent panels. His agreement should be sought to a change of plan at such points or the inclusion of an indent former to provide a clearly defined change of line since this acts as a guide to the concretor and the finisher, and will make the eventual concrete model even crisper.

Chapter 14

MODERN STEEL FORMWORK SYSTEMS

I. DUNKLEY
Managing Director, Datron Gel Ltd

The increase in costs of timber and labour has aroused interest in the UK in formwork systems based on large modular steel panels which have been used in Europe for a number of years. These European systems were first introduced when labour costs escalated and it was necessary to increase productivity as well as reduce overall labour requirements. Concurrently, contractors started to use large capacity tower cranes thus enabling large modular steel formwork systems to become rapidly established (Fig. 14.1).

Large modular steel systems represent a system building concept rather than a formwork system. A complete package which includes formwork scheduling is designed for each individual project. Standard components are also available, but these are only used in the context of the whole building. Formwork made of timber or small steel-framed steel panels represent small mosaic pieces which have to be assembled with a large number of man hours in order to produce an entire building. In contrast, the number of components in a large modular system are very much reduced with consequent savings in labour and time in completing the whole. Quite obviously, the mosaic pieces, whether large or small, must be interchangeable and small panels of timber or steel are, quite clearly, more suitable for projects where there is little repetition. With repetition, large steel panels become economic and by careful design and selection of components, this repetition can be achieved more easily than is often realised.

Increased productivity coupled with labour cost savings are not the only advantages of these systems; high constructional accuracy,

Modern Steel Formwork Systems

surface finish, speed of erection and stripping can often be more important. Certainly, high constructional accuracy is becoming an increasing requirement when prefabricated components have to be installed into the *in situ* building frame.

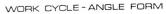

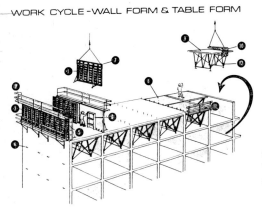

1. **WALLFORM**, crane lifted to next position. Module lengths of .9, 1.8 and 2.7 m.
2. **POURING WALKWAY**, attached to wallform used here as gableform.
3. **GABLE END WORKING PLATFORM** also supports gableform.
4. **BRACKETS** cast in previous slab, support working platform.
5. **STOP ENDS**, fixed, hinged or sliding.
6. **STEEL DOOR INSERTS**, fixed with services.
7. **LIFTING BRACKETS**.
8. **TABLEFORMS** positioned and levelled ready for slab to be poured.
9. **TABLEFORM** being repositioned.
10. **LIFTING JIB**.
11. **WORKING PLATFORM**, stop end, and guard rail.
12. **SCREW JACK** feet.

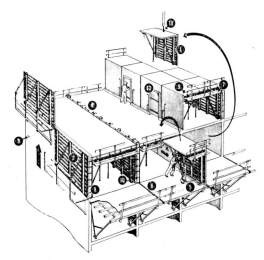

1. **ANGLE FORM**. Module lengths of 2.7 m (3M) with fit-up lengths. Complete with staying and locking devices.
2. **GABLE FORM** with platform, to be lifted as a unit.
3. **U-BOLT**, to be cast in for gable form with platform.
4. **LOW-FORM**. 2 revolving side pieces with cotter pin. The side pieces form a unit.
5. **CARRIAGE** with 3 wheels, for transport of the angles to the working platform.
6. **WORKING PLATFORM** on brackets in underlying cell. One lift by crane for each cell and platform.
7. **PASSAGE BRIDGE** on brackets fixed to the outer angles, follows the form on transport. Rail poles collapsible outwards.
8. **STOP END for wall** hinged in both directions.
9. **STOP END for floor-slab** collapsible, to be locked by 2 cotter pins, the extension piece between stop ends for floor-slab is removable.
10. **ADJUSTMENT SCREW** for raising and lowering the angle form.
11. **SINGLE-POINT LIFT**. Eyebolt with wedge-locking. The bolt hole is placed at the point of balance of the angle. Alternatively a jib can be used.
12. **DOOR INSERT** of steel, tapered 1:10 each side for clearance.

FIG. 14.1. Alternative methods of forming cell structures with large steel panels.

The less obvious, but important advantages of these systems are related to the reduction in the time taken to install services and to finish the building. Fixing holes for cast-in services such as conduits, etc., can be pre-drilled to provide former location thus dictating both position and length of these items which enable prefabrication techniques to be adopted for pipework and services. As most systems are designed on a modular basis, rearrangement of the modules will still leave the fixing holes in the appropriate position on the building grid for use if required. Inserts for door openings, windows, ducts and such like can be positioned in exactly the same way. A high surface finish is possible, large steel panels with carefully designed joints between panels enable decoration to take place without any need for prior plastering with consequent labour and time saving as well as the elimination of wet trades.

Systems of this type can be used for a wide range of buildings and are not restricted to use in housing developments based on the multiple cell principle. Some systems are restricted to housing and apartment construction whilst others have application throughout the building and civil engineering field where repetition occurs:

    Retaining walls
    Culverts
    Bridges
    Underpasses
    Central cores
    Structural floors
    Slab floors

14.1 WALLFORM

As shown in Fig. 14.2, the essential component of a typical Wallform system is the panel and strongback. The panels are generally storey height in modular widths, that illustrated being 900 mm up to a maximum 2·7 m. These panels can be joined together with locking clips, Fig. 14.3, so that a complete Wallform can be pre-assembled prior to being crane lifted into position. The assembled Wallform, complete with props and pouring walkway brackets etc., can be of any length subject only to crane capacity and other handling considerations. After location all inserts for doors, openings and fittings for services and such like can be fixed to pre-drilled holes which will automatically lie on a grid if the building is dimensionally co-ordinated.

Fig. 14.2. Crane lifting of large assemblies considerably increases output rates.

In order to ensure maximum re-use the width of panels used to produce a given wall length will be selected with reference to the building as a whole. The economies of such systems are further improved by the use of strongbacks and all other accessories with a site constructed timber face thus retaining both simple flexibility and the operating economics of the sophisticated system.

Systems such as that illustrated have been studied in use on sites in Sweden by the Swedish Building Institute and compared with more traditional arrangements. Table 14.1 shows the results of this

TABLE 14.1

Productivity comparison of different wall formwork methods

	Work rate (man hr/m^2 cast area)	Output (m^2/hr cast area)
Traditional timber forms	0·48 hr	4·2 m^2 with two men
Small steel panels	0·35 hr	5·5 m^2 with two men
Large plywood elements	0·123 hr	24 m^2 with three men ⎱ incl. crane
Large steel panel systems	0·114 hr	26 m^2 with three men ⎰ driver

comparison and underlines the considerable labour savings that can accrue with these systems. When it is realised that labour saving is not the only advantage, it will be accepted that the use of systems of this type must inevitably increase.

FIG. 14.3. Patent locking devices speed assembly of form for crane lifting.

14.2 TABLEFORM

The use of tableform is now well established in the UK although these are generally derived from scaffolding or support systems rather than having being designed as a specific method of forming slabs. Figure 14.4 shows a typical steel top tableform being moved to a new position by means of a 'C' beam which provides a quick method of removing tables without damage to the walls or previously cast slab. For smaller projects the base frame can be used with a site constructed timber face without detracting from the low operating time and ease of handling.

An alternative method of forming slabs between cross walls is represented by a Danish system which is supported on brackets fixed to the previously cast walls. The slab form is lifted by crane into

FIG. 14.4. Steel or timber topped tableform designed for crane lifting. Note locking devices to join adjacent tables.

position and adjusted by jacks. Stripping is accomplished by lowering the form on to the brackets and withdrawing it by means of rollers which are incorporated in the brackets. This method provides a completely clear span although difficulty will be experienced in achieving a continuous support throughout the striking operation.

Tableform when used in large suspended plate floors with columns can be provided with a wide range of stripping mechanisms ranging from screw jacks to hydraulic or pneumatic mechanisms. The method chosen will depend upon both the variation in soffit height in the building and the stripping height that must be obtained. Individual tables can be joined together by means of locking devices to cover wide spans, and inter-column beams or slab section, ring beams, etc. can be incorporated in the tableform for simultaneous casting, with the slab. Continuous propping of the slab can be achieved during table removal by the provision of propping strips between tables or if the specification allows, propping of part of the table area before other tables are lowered to their stripping position.

Output rates for large steel tableform have also been compared with other systems and Table 14.2 shows the results.

TABLE 14.2

Comparison of slab forming method

	Work rate (man hr/m² cast area)	Output (m²/hr cast area)
Traditional timber	0·25 hr	8 m² with two men
Large panel system	0·073 hr	27 m² with two men

In recent years rib and waffle decks have become widely used in the construction of hospitals, car parks, industrial premises and other buildings with large floor spans. Moulds for slabs for these structures have traditionally been supported on timber formwork and props on scaffolding derived systems. The development of tableform frames with deep stripping mechanisms enables these pans to be permanently mounted on the base frame so that the structural slab can be constructed with as much ease as a plate floor. Figure 14.5 shows the table form used on a large car park project in Gothenburg utilising pneumatic stripping mechanisms. On this project an output of nearly 14 m² per hour was achieved with a three-man crew for a cycle of strip, renovate and erect.

Structural floor systems which feature consumable moulds, or moulds that cannot be permanently mounted on a base frame can be supported on standard table bases provided with a timber frame top when again the slab can be treated in exactly the same way as if it were flat. Similarly inter-column beams can be integrated with table systems and Fig. 14.6 shows a table which forms beam sides. Beam soffits can be formed traditionally, or if the width is great enough, by special form mounted on base frames.

FIG. 14.5. Pneumatically adjustable tables with permanent fibreglass coffers.

FIG. 14.6. Steel topped tableform incorporating edge bottom sides.

14.3 TUNNEL FORMS

Tunnel forms have become widely used in Europe and elsewhere as an extremely fast method of erecting multiple cell structures. Two alternative system types are in use, the complete tunnel comprising two wall faces and interconnecting slab form, and a tunnel comprising two inverted 'L' shaped angles that are locked together along the cell centre line after the formwork is located into position. At first sight the complete tunnel sections would appear to have advantages,

FIG. 14.7. A narrow section of angleform being crane-lifted from a working platform. Alternatively, angles can be lifted by 'C' beam from within the completed cell.

as they do have in certain applications, although the more complicated stripping mechanism and the difficulty in providing continuous propping often outweigh these.

Figure 14.7 shows a large steel panel system section being repositioned on part of a development for 1000 houses in Sweden. The angle sections are erected as a series of adjacent tunnels which enable the floor and wall slab to be cast in one operation. The system is completed by gable end wallform, low forms that are used to locate and cast the kicker on which the next erection of angleform is located and a three-wheeled trolley for the repositioning of the angleforms. Vertical movement of the sections is by means of a 'C' beam which can remove the form from within the cell or by a sling in which case the section is rolled on to a cantilevered working plat-

FIG. 14.8. Angleform on special trolley for horizontal movement.

form as shown in Fig. 14.8. Typical productivity is 0·10 man hours per m², or 30 m² per hour with three men.

A particular advantage of the angleform system is that all formwork loads are carried on the cross walls of the floor below and no loads are applied to the floor slab other than during the erection cycle. The props that are used at that time are hinged so they can be swung clear thus providing complete access to the inside of the form.

14.4 SPECIAL FORMS

Central cores for lift shafts, staircases and such like can be produced from wallform together with special items to enable a complete custom form to be produced for a given building. A typical form is illustrated in Fig. 14.9. Individual components for these core forms are selected to ensure the maximum re-use throughout the building and this often enables four or five completely different cores to be produced from the same form elements. Special stripping devices are incorporated on the inner form and returns on the outer form, and inserts provided for door openings and location of other structures.

A systems approach can also be adopted in the case of columns and Fig. 14.10 shows steel column forms ganged together to reduce the time taken in setting out façade columns. Forms of this type can be adjustable, both in column size, as can standard steel columns, and in pitch, thus achieving greater cost economies.

Special formwork for entire buildings can also be designed using standard produce for design technique flow applicable. Figure 14.11 shows a hospital in Copenhagen where special formwork was used for staircase cores and the internal faces of walls, the external faces being precast and supported by special strongbacks.

FIG. 14.9. Stair and liftshaft core form utilising standard and special components.

If large continuous vertical surfaces are to be cast, sliding or clip form can be utilised and Fig. 14.12 shows such a system with manual winching of the form face. After each form lift, the lifting frame complete with walkways is itself jacked to the new position.

Fig. 14.10. Ganged façade column form.

Fig. 14.11. Special form for composite precast and *in situ* walls.

FIG. 14.12. Manually jacked sliding form.

14.5 PRECASTING

Although this subject is beyond the scope of this chapter, mention must be made of precast concrete in connection with large steel panel formwork. As stated earlier these systems represent more of a system building concept than pure formwork and they are more similar to precasting moulds than they are to traditional formwork. This fact is illustrated by the ease with which standard components can be incorporated into precast moulds for site use and the fact that manufacturers of sophisticated steel systems invariably include a range of precast moulds in their product range.

14.6 GENERAL

As steel formwork of the type mentioned in this chapter becomes more firmly established throughout the world, it is interesting to examine the various reasons for adoption in different countries. In Sweden, despite the availability of timber, it was necessary to reduce the labour requirements of formwork without increasing the rate of completion of the building. This enabled a small formwork team with a relatively small amount of formwork to offer considerable economies on large-scale building developments where rapid completion was not essential. In the UK the need is both to reduce labour costs and also increase the rate of completion which necessitates a larger surface area of formwork.

In the USA high labour and timber costs have, in some States, so altered the competitive situation that steel-framed buildings are becoming increasingly popular in medium rise applications. Steel form offers a means of counteracting this trend. Finally, in South Africa where all timber has to be imported and there is a serious shortage of skilled labour, steel forms reduce the skill requirements and the actual labour involved.

Chapter 15

PLASTICS AS A MOULD MATERIAL

PETER J. OWEN
Director, Bondaglass-Voss Ltd

Plastics are easy and straightforward to use and a variety of plastics are available for use in making moulds for concrete.

The use of plastics involves the same basic techniques that are in everyday use in a concrete company, namely making one material from many and 'batch mixing' in order to do so.

15.1 WHAT DOES A PLASTICS MOULD MATERIAL OFFER?

This list is not necessarily in order of importance since companies have different priorities:

Ability to conform to any shape since it is used in either its liquid state or can be moulded by heat.

Great rigidity and strength and comparative lightness.

Resistance to chemical attack and weathering.

Uniformity of concrete quality and colour.

High use capability if handled with reasonable care.

Specialised equipment is unnecessary in most cases and, therefore, the material can be used within the normal framework of the concrete company's mould shop.

On the aspect of price it is always necessary to consider the use to which the mould is being put together with the number of casts that are required from it if the 'true' price is to be established. It would be reasonably correct to say that plastics are competitively priced as compared with other mould materials when the factors above are taken into account. Plastic materials are not a universal solution for the whole range of moulds required. There are, of course, some that are far better made from other materials.

It is important when considering whether or not to use plastics that a general appreciation is made of the raw material state, capabilities, advantages and limitations of each type of plastic.

15.2 GLASS REINFORCED PLASTICS (grp)

Working conditions and equipment. One of the materials used in the production of polyester resins is styrene. This is an irritant, although non-toxic, and therefore it is necessary that any area in which polyester resin is being used is well ventilated. The most suitable working temperature is from 14°C to 20°C.

It is advised in order to mix the polyester resin easily and quickly with the catalyst (hardener), that an electric drill (plus/minus 1000 rpm) together with a mixing propeller is used. This is the only equipment that is required in order to mix resins although, of course, mixing can be done by hand using a wooden stick.

The glassfibre mat itself is best mounted on a spindle at one end of a long cutting bench since this makes the measurement of the required amount of glassfibre much easier and more controlled. A sharp knife or pair of scissors is quite suitable for cutting the glassfibre.

15.3 STORAGE OF MATERIALS

The polyester resin should be stored in a dry and cool place with a temperature of around 20°C. In these conditions the resin should have a shelf life of between 6 to 9 months. The accelerator (where pre-accelerated resins are not in use) and catalyst should be stored away from the resin and since both these are highly inflammable care should be taken in their storage. Neither should the accelerator or catalyst be stored in close proximity to each other since if mixed together, undiluted by the resin, there is a possibility of self-combustion and even an explosion.

With regard to solvents for liquid polyester resin there are two types. The first is acetone which is highly inflammable and thus also needs to be stored carefully and the other is methylene chloride which is toxic but non-inflammable and here again care should be exercised.

15.4 TOOLS AND EQUIPMENT

The tools required for laminating glass reinforced plastic are firm bristled brushes and metal disc rollers. It is preferable to use white bristled brushes and the normal size is either a 50 or a 75 mm brush. Lambswool rollers can also be used for applying the resin and a 125 mm size is found to be the most practical.

Metal disc rollers are used for excluding the air from the laminate and are supplied in a variety of diameters and lengths. The size of roller used will depend on the nature of the laminate being moulded. Typical sizes are 50 mm or 100 mm long by 40 mm in diameter. There is also a roller 75 mm long by 6 mm in diameter which is particularly suited for rolling mat in corners and other curved sections.

The task of trimming glassfibre mat after it has been laminated with polyester resin depends on the stage at which it is trimmed. While it is still in a 'leather like' state it can be trimmed easily with a pair of scissors or a sharp knife. When it has hardened or cured the trimming can only be done with either a special glassfibre trimmer, a sanding disc or a hacksaw. Where further smoothing is required wet and dry paper can be used and the grade varies according to the degree of smoothness required.

15.5 MEASURING EQUIPMENT

In the early stages of using polyester resins it is advisable to have accurate measuring equipment. This is required because the 'pot life' of polyester resin is controlled by the amount of accelerator (if using an unaccelerated resin) and catalyst added to the resin. It is easiest to have metric measuring equipment since it makes calculations of percentages simpler.

It is also advisable to have a glass measuring cylinder marked in cubic centimetres for the measurement of catalyst.

15.6 POLYESTER RESINS

There are many types of polyester resin but the two most commonly used in concrete mould manufacture are:

Gelcoat resin
Standard polyester laminating resin

Gelcoat resins are more thixotropic than normal laminating resins and give the hard smooth surface to the mould. This is the pigmented layer in any laminate and protects the glassfibre from alkali and water attack. Gelcoat is applied with a brush and it is well to estimate on the basis of 600 g gelcoat resin per m^2 of mould surface area.

Polyester resins are supplied in two qualities. One is isophthalic and the other orthophthalic. Standard laminating resin is of the latter type and it is generally unnecessary to use the higher quality isophthalic resin. The only time in which this would be used is where better heat and distortion resistances are required, 105°C compared with 80°C for the standard laminating resin.

It is advisable to use a pre-accelerated thixotropic resin for 'lay up' since this is easier to use and eliminates the need to calculate and measure the quantity of accelerator to be added leaving only the addition of the catalyst or hardener.

15.7 CATALYST (HARDENER)

The catalyst is also referred to as a hardener. The normal catalyst for polyester resin is a MEKP (methyl ethyl ketone peroxide). The quantity added to the resin controls the 'pot life' or working time as well as the curing time. At the working temperature previously stated it is normal to add 1% of catalyst to the resin since this will give a pot life of some 20 minutes. By adding more catalyst the working time is reduced and by adding less the 'pot life' is increased as is the curing time. The action of the catalyst is to promote a chemical reaction and during this time heat is caused. For this reason it is practical if required to shorten the curing time for a laminate by putting it into a room at a higher temperature or using some other form of mechanical heat.

As said earlier it is preferable to use pre-accelerated resins but if a non-accelerated resin is bought it will be necessary to add the acclerator, usually cobalt naphthenate, to the resin as well as the catalyst. Here the manufacturer's specifications should be followed and in no event should the accelerator and catalyst be mixed together. Always add one of the two to the resin and mix it in well before adding the other. The reason is that if the two are mixed together an explosion can result.

Catalyst is caustic and care should be taken to prevent it getting on to the skin. If this does happen then the catalyst should simply be washed off with plenty of water. It is very important, however, to avoid getting it into the eyes. In the case of this happening it is vital that the eye is rinsed either with an eye bath or simply by pouring water into it. In all cases the eye should be washed immediately, consulting a doctor as soon as possible afterwards.

15.8 GLASSFIBRE

There are basically three types of glassfibre. *Surface tissue* is a randomly woven mat of very fine strands of glassfibre. It is used to strengthen the gelcoat and prevent any of the mat from showing through the gelcoat. *Woven cloth* is similar to any woven fabric and is available in several weights and weaves. The most commonly used is *glassfibre mat* which is woven like surface tissue but is thicker and is normally supplied in three weights 300, 450 and 600 grammes/m^2. The normal weight to use, since it provides both the highest strength and easiest workability, is the 450 grammes/m^2 mat.

The number of layers of glassfibre used depends on the size of the mould itself. For a normal mould, three layers of 450 grammes glass mat are suggested making a total weight of 1350 grammes/m^2 of glass. The amount of resin normally required for one layer of 450 grammes mat is between 900 and 1350 grammes/m^2.

15.9 THE MASTER MOULD

The master mould is the most important and costly single item. If the gelcoat surface is required on the outside of the casting then it is necessary to use a negative mould. Thus it may be necessary to first build a positive 'plug' from which a glassfibre negative can be taken. The surface of the plug would be highly finished and treated with release agents after which the gelcoat would be applied followed by the glassfibre and polyester resin. If on the other hand a hard shiny surface is required on the inside a positive plug would be used.

Whichever method is employed it is as well to remember that the casting taken from a mould can be no better than the master and will reflect every imperfection in it.

15.10 RELEASE AGENTS

There are several types of proprietary release agents. The most common are either a wax emulsion for porous surfaces such as plaster or wood and a PVA for non-porous moulds such as metal. Generally, however, a non-silicone wax polish is used where repeated mouldings are being laminated.

15.11 MOULD REINFORCEMENT

If the negative or positive mould is going to be used to manufacture many mouldings then it is necessary for this to be stronger than any of the mouldings to be taken from it. Thus more layers of glassfibre mat and polyester resin will have to be laid. In addition it is also advisable to reinforce the moulds by using wood or metal reinforcing bars in strategic places. These can best be bonded on to the laminate using a polyester filler.

15.12 LAMINATING

The surface of the mould is first treated with a wax release agent which is polished, to a high finish. If the gelcoat is to be pigmented the whole quantity required for the work should be pigmented at one time so as to ensure a standard colour. No more than 10% of colour paste should be added by weight and it is preferable to add as little as possible. The amount required will depend on the colour. Once this has been done the measured quantity of catalyst is added to the gelcoat, stirred in well and brushed as evenly as possible across the entire surface of the mould. The gelcoat is then allowed to cure until it is sufficiently rigid to prevent the glassfibre mat from going through to the mould surface. It is normally still tacky.

When this stage is reached a coat of catalysed polyester resin should be applied to the first area to be laminated. By doing this it is possible to achieve quicker penetration of the glassfibre mat by the polyester resin because, having applied this initial coat of resin, the glassfibre mat is placed on top and a further coat of resin is applied using a stippling action with the brush. The mat is, therefore, impregnated from both sides. During the course of the next two or three minutes the individual glassfibre strands will almost disappear into the resin becoming transparent as the binder which holds the mat together dissolves. When this stage is reached it will be necessary to use the metal disc roller to force any entrapped air to the surface thus ensuring a good quality laminate. The presence of trapped air can be recognised by white coloured patches in the glassfibre mat.

This rolling action will, of course, not only force air to the surface but also resin though this will, when the rolling action ceases, normally sink back into the mat. Any excess resin should be removed by the brush since this does not contribute in any significant way to an increase in strength. The sequence is normally as follows:

1. Release agent is applied and the mould polished
2. Gelcoat applied
3. Surface tissue is applied
4. Glass mat/roving stippled and rolled
5. Glass mat/roving stippled and rolled
6. Glass mat/roving stippled and rolled
7. Reinforcement as required

The use of glass cloth has not been included here since this would not be necessary for concrete moulds.

15.13 USAGE

Glassfibre reinforced plastic moulds have a high potential number of uses. However, the actual use obtained from moulds does depend on the care with which they are handled. Some maintenance is normally required.

15.14 RELEASE OILS

Polyester resin is always subject to saponification and thus the use of release oils is recommended.

Advantages
1. High potential use
2. Possible to make complicated shapes
3. Seamless mouldings can be made
4. Relatively inexpensive compared with metal
5. Relatively easy to prepare
6. No capital investment required in equipment

Limitations
1. Concrete tends to craze
2. Concrete tends to mottle

These relatively minor limitations can be eliminated where the concrete is lightly ground, grit blasted or acid etched at a suitable time after striking.

15.15 THERMOPLASTICS MOULDS

These are a very useful group of materials since they can be moulded into complex shapes by applying heat and vacuum. There are three basic types of materials:

PVC—Polyvinylchloride
ABS—Acrylonitrile-butadiene-styrene
Polystyrene

They are supplied in sheet form in various sizes and thicknesses and normally have a high gloss surface but can be obtained in a matt finish. None of these is really capable of dealing with undercuts although the sheet does have a degree of flexibility and resilience.

The precaster is unlikely to have the necessary vacuum forming machinery but there are many suppliers who are able to construct the necessary mould and form the sheet.

A former or master is required over which the plastic sheet is drawn. This can be made from most solid materials but for long runs metal moulds are best. Besides the shape all that is required is for 1 mm holes to be drilled through the mould at about 100 mm centres. While there is great flexibility in designing it is as well to consult a vacuum moulding company on the practicability of the design at an early stage. Normal concrete 'draw' should be allowed in the design.

15.16 THE METHOD

The 'plug' is placed in the vacuum former where the plastic sheet is heated and then drawn by vacuum over the mould. The draw should be kept to a maximum of some 40 mm otherwise arrises and rebates tend to become thin and the mould will tend to crack. They are virtually impossible to repair.

The production time is related to the size and thickness of the sheet. In the case of thin sheets it is seconds, while for thicker sheets it can be several minutes. The completed mould can be trimmed to size with a guillotine or a sharp knife. It is then preferably stuck to the formwork using an adhesive recommended by the manufacturer or it can be screwed or nailed down. Reinforcement may be required, if the mould is a large one, and this can be provided by filling the hollows with plaster or one of the expanding foams such as polyurethane.

The size of an individual vacuum moulded sheet is limited to a maximum of 2440×1220 mm although a number of sheets may be welded together to provide larger areas.

15.17 CONCRETE

Since the surface of the sheet is highly finished the resulting concrete is similar but this weathers to an egg-shell finish which is retained. It is best to use a slightly over sanded mix.

There is no need to use release oils and the concrete should be left for at least 48 hours in the mould to ensure a sufficiently hard surface so that it will withstand the suction of the plastic. When several moulds are being joined together in order to produce a larger casting this can be done using tape or by heat welding. This latter method is not an easy operation and should only be carried out by someone experienced.

Vacuum formed plastic moulds should be carefully handled since they are easily broken and cracked. With careful use up to 20 concrete castings are possible from the same mould. It is, of course, a simple matter, having constructed the master, to produce replacement moulds.

15.18 TWO-PART POLYURETHANES

A group of materials based on one of the most flexible of plastics—polyurethanes. They have an ability to intermix chemically with a wide range of other materials and this enables them to be used in many ways.

The one-component polyurethane has been used for some years as a means of achieving easier striking and uniformity of concrete by sealing the surface of the mould.

15.19 AS A MOULD MATERIAL

Many moulds and formers (checkouts) are used in casting concrete often complex in shape; these have been a problem to the precaster and mould maker for years. The main problem has been that the material from which they have to be manufactured has normally been a solid material and perhaps one that can only be produced by a specialist company, for example aluminium or steel. Coupled with the choice of material has been the consideration of accuracy that has become so important. Yet another factor has been the re-use potential while maintaining both accuracy and the uniformity of colour of the concrete surface.

A recent introduction into this country has been a two-part polyurethane. Used for some years on the continent in a variety of roles, the material provides a valuable contribution to those features already provided by timber, steel, glass reinforced plastics, and other casting resins available on the mould materials' market.

Polyurethane is extremely tough and resistant to abrasion and in the formulations available is resistant to moisture, alkali, and weak acids. The material is virtually self-releasing and suitably handled will provide upwards of 100 uses in casting.

Because hardened concrete does not adhere to it little force is required to remove the polyurethane formers; this reduces damage to the concrete. Since there is no absorbtion of water the colour of the surface finish of the concrete is generally quite uniform.

Using simply skilled labour, the polyurethane compound can be cast into faced plywood and timber moulds, and provided the normal skills of mould making have been employed the resultant formers will comprise neat, dense material with cleanly-defined arisses and features liable to minimal shrinkage and distortion. Several degrees of elasticity can be produced, thus widening the range of applications, and the material can be used in combination with other materials such as ply, glass reinforced plastic, and concrete in such a way that the best advantage can be taken from the natural properties of these materials; namely rigidity and flexibility, workability and density.

The more flexible formulations are finding an ever growing application in providing moulds for decorative relief and ornamental concrete. In addition it is now common practice to cast tiles into concrete to provide a decorative finish. Until these flexible materials were introduced it was general to use wooden moulds for this purpose. The moulds were very liable to damage on striking and grout ingress was a problem given that the tiles varied in size. A flexible plastic mould solves these problems. It has a high usage factor and there is little 'making good' required.

The material is liquid and conforms to the shape of mould into which it is poured. It is 'cold cured' and hardens at temperatures as low as 0°C.

The adhesion of polyurethane to most materials (plastic excluded) is good and thus fittings can be incorporated at the pouring stage thus obviating any further drilling or tapping for fixing the checkouts into moulds or forms.

As with concrete, a 24-hour casting cycle can be achieved using polyurethane, and where a number of formers are required simply fabricated faced plywood moulds of gang formation will provide a ready supply of formers for incorporation into the production moulds as required.

In cases where small numbers are required, individual faced ply moulds will provide the means of indent production with very accurate formers resulting. The process of manufacture is one which can be carried out in the mould shop using the existing trade skills, and the precaster retains the initiative in progressing the manufacture of his formers.

The chemicals must be used in a well-ventilated workplace and particular care is necessary with one component which is toxic when liquid, the volatility of the material is low and provided the working instructions are followed the permissible level of fumes in the air will not be exceeded during the mixing or hardening process.

The material can be used in a solid state, or formers can be manufactured incorporating recesses where necessary. Reinforcing steel can be built into the formers to stiffen large units and ensure that grout infiltration at the edges of the former is minimised. Fixing to the main mould can be by screwing or, in the case of the larger checkouts, by bolting in such a way that the fixing bolts double as extractor bolts at the time of de-moulding.

Carefully used, with attention to lead and draw, the material should be extremely useful to the precaster. The mixing process is simple and the mould designer and pattern shop supervisor, well known for their resourcefulness, can undoubtedly develop the techniques to suit the particular requirements of their company's production methods.

15.20 OTHER PLASTICS MOULD PRODUCTS

The polyurethane material is a relatively new introduction. There are other plastic based compounds available that achieve the same result. One of the better known materials is a vinyl rubber compound. This is not a cold curing material but one that requires melting, and is a plasticised PVC. Normally these linings are supported in rigid pallets to avoid distortion both when they are being cast and also when the concrete is being poured.

The master mould against which the molten material is poured can be made as in the previous case from almost any material. An advantage of this material is that it can be used over and over again simply by remelting it, although the heating process is often rather a tedious business.

15.21 EXPANDED PLASTICS

For producing both shallow and deep relief concrete on a 'one off' basis the pre-expanded plastics provide an excellent material.

For the precaster there are only two foams that can be used with any degree of ease. These are polyurethane and polystyrene since they are supplied as sheets of expanded foam and are ready to use.

Both these materials are reasonably cheap but it should be remembered that in the main the former can only be used once since in order to remove them from the concrete it is usually necessary to destroy them. On the shallower moulds it is sometimes possible to use a release oil and thereby obtain more than one casting.

15.22 MATERIALS

Sheets of foam are, of course, available in a variety of standard sizes, thicknesses and densities. However, it is possible to have the sheet cut to the required dimensions by the manufacturer since the foam is sliced off a large block.

Of the two, polyurethane is the most stable because it is unaffected by solvents, acids or alkalis whereas polystyrene literally disappears if it comes into contact with any of the more common solvents. This can, in some cases, be an advantage when it comes to removing the foam from the concrete.

Generally precasters use the higher density foams since these withstand the weight of the concrete. Even these foams require additional reinforcement in the way of a frame to support the mould. While it is possible to specify a required foam density it would be preferable for the precast company to ask the advice of a foam manufacturer to recommend the most suitable for the purpose.

The pattern is normally carved into the sheet using one of several methods:

    Urethane and styrene—by heat
    Styrene only—using solvents carefully
    Urethane and styrene—using hand tools
    Urethane—by bonding

Both types of foam are reasonably easy to carve with a knife or other sharp cutting tool. In the case of polystyrene a solvent such as petrol can be used to dissolve the foam but this particular method needs either absolute control and knowledge of the dissolving rate or a freedom of design that can accept any irregularities. Polystyrene foam is very difficult to repair satisfactorily since most repair materials contain some solvent.

Heat will melt both materials and, therefore, can be used to form and shape the foam. While this method is easier to control it can still produce some unexpected results! Of all these methods the knife is the most direct and controllable and is the method normally used.

The sheets have to be treated carefully since even the higher density foams crush easily. The main difference between the two in handling is that 'urethane' tends to 'dust' while 'styrene' crumbles.

It is, of course, possible to repair polyurethane since it is not susceptible to solvent attack and a polyester filler will bond to the foam extremely strongly. In fact it is possible to obtain a re-usable mould by coating the foam pattern with polyester resin and glass-fibre. A surface coating can be given to polystyrene by using the solvent free polyurethane referred to in another section. This can be applied with a brush, will not affect the styrene, and will provide a smooth, non-porous casting surface.

When panels have to be joined then polyester filler can be used in the case of urethane but either tape or a solvent free adhesive would be required for polystyrene.

The removal of the foam from the concrete surface is either done by hand with a wire brush or in the case of styrene a solvent can be used. It is also possible to grit blast the foam away.

While the precast industry has not generally used plastics as a mould material it is becoming more aware of their possibilities. This has been brought about by several factors such as the demand for complex shapes, relief concrete and better standards of finish. In addition the conventional materials have risen in cost and this has prompted further examination of suitable alternatives.

Plastics of the types available for mould forming have also suffered from a feeling that they are complicated materials, difficult to use and unpredictable. They are none of these and when used properly and with reasonable attention to instructions are as simple or difficult as concrete.

The purpose of the information contained here is to provide a general appreciation of the materials available and their respective uses and it has not, in the main, been to detail the method of use, since this can be best given by the supplier in discussion with the concrete producer or contractor.

Again it is of no use having an appreciation of plastics unless there is also a willingness to use the materials themselves at least on a trial basis. Unless this is done it will be impossible for the full potential of the materials included here to be evaluated.

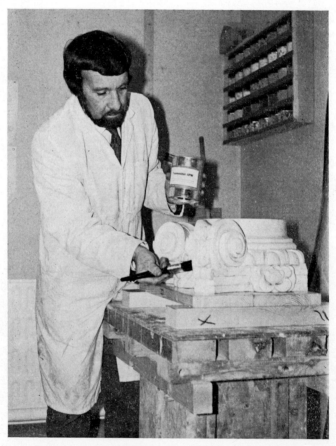

Fig. 15.1. Release agent being applied to plaster master prior to producing polyurethane mould.

Fig. 15.2. Reconstructed stone cast from flexible polyurethane moulds.

Fig. 15.3. Master unit positioned prior to pairing of polyurethane mould.

Fig. 15.4. Plastic coffered floor formers.

Chapter 16

SPECIAL STEEL FORMWORK—
A CASE STUDY

P. R. LUCKETT
Technical Director, Stelmo Ltd

A contractor who secures a contract to build a large concrete structure is faced with a great number of problems. He must, as a primary duty to his shareholders, use the most economical methods. This should not be difficult as he has wide experience and has in all probability executed similar jobs from which he will gain valuable information; but this is rather like making the same journey that one made sometime ago and navigating from established landmarks, the only problem is that someone has been moving the landmarks slightly so that the course must be changed somewhat if the shortest and most economical route is to be taken.

The landmarks a contractor will use are cost and availability of material. It can be cynically assumed that cost varies in inverse ratio to availability! It is highly unlikely that a commodity in short supply will be reasonably priced. The commodity can be either skilled carpenters or high quality plywood sheet. The point being made here is that because a contractor has always made his own forms it is not necessarily the case that it will always be economical for him to do so. He must be constantly aware of changes taking place.

Other factors are the ever increasing cost of any type of labour making more attractive the utilisation of possibly expensive plant. This can lead to hire rather than purchase of these capital items which, again in turn, as the hire will almost inevitably be on a time-rate basis, lead to intensive and continual use so that it may be returned to the owner and removed from hire as quickly as possible.

When this philosophy is applied to the placing and forming of concrete in any type of repetitious situation the tendency is to think of the formwork in very large pieces so that crane occupation will be reduced. In this way far less lifts are required and the crane will be released for work elsewhere. Large sections of formwork also mean less joints with a consequently lower labour content in stripping and re-assembly. Less joints also mean much more accurate work as there are far fewer movable surfaces to be disturbed. There are naturally problems which can be encountered with large sections; unwieldiness and sheer physical weight, coupled with handling in high winds, must make a contractor think about having adequate means of coping with these situations.

Accepting that large forms are an economic solution to be considered, the question of material arises. One must automatically gravitate to steel as, unless very elaborate deep trusses are used, large spans cannot be undertaken in timber. The vulnerability of timber makes its utilisation in arduous conditions a matter for careful thought. As an imported material in the UK it is also subject to the various international price fluctuations which can happen. Steel, of course, being a raw material which is produced in this country from ore excavated here and having its price controlled by the Government is not subject to such a world inflationary trend; although we have recently seen a quite marked increase in the past few months, it is not so severe as the world price rise in timber. While aluminium has a real advantage from the weight aspect, its cost does really remove it from use as a structural material.

The contractor is then considering a special steel form and he must decide whether to make it himself or approach a specialist manufacturer. He will save money if he makes it himself, but it is very doubtful whether he will have the skilled staff to do this type of work. He will then go to the specialist, specify his needs and in due course receive an offer of a service. Then will be the time to carefully study all the aspects of the job to make sure that this is the correct solution. The total cost of the steel form will probably shock him if he is unfamiliar with this type of work, but for this he should be getting a total service involving him only in the provision of labour to operate this form. The capital cost must invariably be written off against the job as it is most unusual for the form to have any residual value due to the designers always making consecutive structures of a slightly different shape and the difficulties of storage and re-marketing.

It may be that study of a particular contract in detail involving the use of large steel forms will help to clarify their use in the readers' mind. The contract chosen is the Ouse Bridge, which spans the Yorkshire Ouse near Goole and is part of the motorway scheme for that area. The main contractor is Messrs. Costains Ltd who commissioned the consulting engineers Maunsell and Partners to design an alternative solution based on a Redpath Dorman Long Steel Deck. The consultants for the whole scheme are Scott, Wilson, Kirkpatrick and Partners and the client the Department of the Environment.

FIG. 16.1. Crosshead form erected at site. Access scaffolding is an integral part of the formwork design.

Figure 16.1 shows the columns that are to be parallel which is an immediate advantage as a standard height form can be used with simple underfilling when a shorter column is required. To expand on this point, many support columns are tapered in elevation which is usually a visual requirement as, structurally, it does little towards the strength of the configuration. This in itself would be quite innocuous were all the base dimensions of the same size and the top dimensions varied depending upon the height. Then the form could again be underfilled to make columns of differing height. However, when a tapered column is used, the top is invariably made the constant and the base dimensions varied to suit the height, that is, the taller the column the smaller the base which is a structural contradiction but is no doubt visually satisfying to those who have time and opportunity to gaze at the structure. When a tapered column has to be cast with a varying height and a constant top dimension the cost and complexity of the form is considerably increased. Height variation must be achieved by removing sections from the bottom of the form which entails joint lines and mating seams. The other solution is to allow the vertical ends to slide within the main side panel, but this can only be done if the column has a rectangular cross-section. It will, therefore, be seen that a parallel column gives a lot less problems in the manufacture and use of a large steel form.

A useful system developed for tall columns is the landing ring principle. If the reader is unaware of this he should imagine a normal column form which has a small additional section of form bolted to the top and the bottom. If one thinks of dimensions the height of the centre section is probably about 8 m and the heights of the top and bottom section about 600 mm giving an overall form height of 9·2 m and a pour height of 8·6 m. After the initial placing the top section of the form is left around the column and the main form and bottom sections are stripped and lowered to the ground. While on the ground for cleaning, the section which was at the bottom is detached and transferred to the top and bolted to the main form. When this assembly is made on the ground the whole of the form can be lifted to the top of the partially completed column and bolted on to the section which was left around the column at the top of the pour. Concrete placing is effected and when the concrete is strong enough the large centre section and bottom which was left around the column are detached and lowered to the ground leaving the last top section around the top of the column where the concrete placing was completed. The process is repeated again and the reader can see that the column is cast in a series of continuous lifts without the form ever leaving the concrete at the interface of the two pours.

The Ouse Bridge employed an extension of this system as can be seen in Fig. 16.2. The main column form had a detachable landing ring which was built so that the jump system could have been used if required. Here, it can be seen that the contractor had, in fact, three forms, two of which were fastened together to continue the pour. This column was not actually poured in one although it appears that way. The form used for the first pour was just left around the column as a convenient storage place until the reinforcement cage was ready and in position for the next column. A point of interest in these forms is that the principle of stiffening a form face with soldiers and walers is still maintained; in this, the walers become heavy ring beams which run right around the column. It should also be noted that there were safety hooped access ladders and a working platform which was boarded out, and that tubular scaffold hand-railings were used by

FIG. 16.2. Column forms re-erected. The lower set of forms on the left-hand column have been left in position and the second lift forms erected on to them.

the contractor. These forms are usually designed for quite high rates of pour.

The crossheads which support the road deck are shown in Fig. 16.2 with the formwork around the beam in the foreground. The traditional approach to a problem of this type would be to support the soffit of the formwork directly from the ground. The ground on which the support system was founded would need to be stabilised in some way so that movement likely to disturb the support structure would not take place. This can be difficult and expensive, especially if the ground is soft and waterlogged.

A considerable saving in foundation preparation can be made if the existing foundations for the columns can be used. This may not be easy if the column is an extension of a rather narrow pile cap, but where a large platform base exists, then there is scope for utilisation. Substantial purpose-made raking props can be based on these foundations and utilised to support the falsework. This avoids ground preparation, but still involves long structural members which carry considerable loads. The logical extension to this is to use the column itself as the support and the most satisfactory structural method of doing this is to pass a beam through the column. This normally involves moving the reinforcement apart at the centre, but this is an alteration which can usually be negotiated with the consultants without too much trouble. The hole for this beam can be seen in the top of the column in the background of Fig. 16.1 and in Fig. 16.2 the feint marks can be seen where the holes have been filled. When the steel beams have been positioned, the erection of the formwork can then continue.

A contractor will want to position his reinforcing cage on a free and open soffit so that he can obtain proper access for his fixers. He could erect within the confines of the complete form if necessary, but again following the earlier edict of large sections with big lifts, he will no doubt prefer to pre-assemble large sections of the cage at ground level and lift them up in one piece to the crosshead form. All this means that the soffit must be in position before the sides, and be able to carry the cage weight. This is achieved by spanning the gap between the columns with two beams which rest on the through beams. These are shown in Fig. 16.4 as the main beams along the underside with cranked up ends. The brackets which carry the walkway can also be seen with the small pockets at the end which carry the handrail.

Figure 16.3, which shows a partial works erection, also illustrates the way that the soffit is raised in the centre. It was proved at the form design stage that the problems associated with following the beam profile on the underside would be far more expensive than using straight support beams and form edge, and lifting the steel soffit on

FIG. 16.3. The crosshead form partially erected in the manufacturer's works showing the stooling to the beam profile.

FIG. 16.4. The crosshead form which shows the support beams and various brackets which support access ways. The combined skin/carcass construction is clearly visible.

permanent stools. These show quite clearly emerging from the part erected sideform.

When the cage has been fully assembled and checked, the sideforms are placed on top of the soffit support cross beams and clamped home, with the special clamps. These are special because they transfer the total load of the structure into the sideforms, suspending the whole soffit with its load of wet concrete. Top cross ties are placed across the two sideforms so that they act in unison and then the assembled form is ready for filling.

Concrete placing is a critical operation in itself as the steadily increasing and varying load must be controlled at all times. Placing must begin over the columns to stabilise the points of maximum load and avoid eccentric loads on the form. This means that the contractor ideally has four faces to work on at the same time—ideal from the point of view of loading but not from that of the concreting foreman who must control the operation that has be completed in a single and continuous cycle!

When the concreting has been completed and the top surface screeded and trowelled, the beam can be left until it has achieved sufficient strength to carry its own mass and superimposed loads. Enlightened engineers now permit the removal of soffit support when the concrete has achieved a specific strength. This solution usually places the responsibility on the contractor in that some specifications give permission to remove supports when the contractor proves to the engineer's satisfaction that the concrete has achieved sufficient strength to carry its load without deformation or detriment to itself.

In designing a steel form the use of many disciplines is brought to bear, none the least experience. Recent collapses of falseworks and partially completed structures have made it necessary to examine the design methods employed. In the absence of anything better BS 449 gives safe and reliable methods which can be used with perfect confidence. The problem is that this specification gives design criteria for permanent structures, and formwork is anything but this. For example the stiffening effect of partially set concrete is completely ignored. There are other anomalies which make it desirable for a special set of design rules for formwork alone to be formulated based on BS 449, but accepting the fundamental differences in usage of the final structure.

The Technical Report published by The Concrete Society and

The Institution of Structural Engineers on *Falsework ... Report of the Joint Committee* is a valuable document and gives engineers concise information on the whole subject in one publication. It has, unfortunately, not been possible for the report to specifically separate formwork as this is not looked upon as a material which directly carries structural loads as opposed to fluid loads of the wet concrete. Conventional soffits usually transmit their load to the supporting falsework at fairly close centres and, therefore, do not become components which figure in too much detail in falsework analysis. Large purpose-made steel forms which obviate the use of falsework must be subject to the same rigorous design, analysis and examination procedure.

The interim report of the Bragg Committee makes some valuable observations on the current situation. The proposal that a single individual should authorise that a particular structure is ready for the concrete to be placed is interesting and, if implemented, would ensure that a competent person was making the stability of falsework his own specific job, hopefully, without the pre-occupations and pressures of a number of conflicting requirements. The submission and approval of calculations is something which should always be carried out. The fear is that all these proposals could involve considerable additional cost. This need not be so, as, without doubt, calculations are prepared and this preparation can be formalised with little extra work. If no calculations are prepared, then the structure is either so simple as to not require them, or, they should have been, and the disciplines called for in these various reports must be implemented.

Large special steel formworks are complex and expensive structures. The design procedure is quite sophisticated and in certain instances has to be placed on a computer for accurate results. This manufacture involves the utilisation of highly skilled labour all of which adds up to a high cost. Although the economics of a job can show that for a reasonably tall crosshead a stressed skin form of this type can be the cheapest method of doing the job, it does seem an economic waste that a form is discarded after it is barely run in after six uses. Why are they not re-used more frequently? One reason could be that a contractor may be unaware of the existence and whereabouts of a form which could be utilised. Because of the specific design techniques used, and the tailoring that comes into one particular job, they are difficult to modify in order that they can be used on a structure of slightly differing dimensions. There does seem to be a very strong case for the standardisation of this type of structure which would immediately result in a form having possibly hundreds of uses leading to an even more efficient design from the labour content angle. There seems to be no good reason why this cannot be done on crosshead beams. The end profile which may be a matter of the whim of the designer could be done in timber as this is a fairly low stressed position. The question of column standardisation becomes more one of aesthetics, and some columns can be found parallel in side elevation with a strictly rectangular cross-section while others are tapered in elevation and have a number of cross-sectional features which other designers may not wish to repeat. With standard highway widths and standard beams in their support structure it is difficult to imagine why crossheads have never been standardised—perhaps nobody has thought of it before but the existence of standards would reduce the amount of work required to design the structure and enable the large sections of formwork to achieve the number of re-uses which would show real final savings on the job.

Chapter 17

SLIPFORM

C. J. WILSHERE
John Laing Design Associates Ltd

17.1 INTRODUCTION

Traditional formwork is a repetitive technical operation. Forms are erected, concrete is placed, it hardens and cures, and the formwork is removed ready for use elsewhere. When this process is made continuous, we have slipform, better known in the United Kingdom as sliding formwork. A method can be successful technically but it will not be adopted unless it offers a financial advantage. In this case there are several factors which contribute to economy. It is possible to construct the structure in a much shorter time than by conventional means. This often eliminates a bottleneck, which significantly reduces the overall site programme. It is usual to slide without traditional scaffolding. There is normally a complete absence of the weakness caused by construction joints and the process is sufficiently unusual to attract much interest and enthusiasm from those who have to carry it out, and the overall cost to the contract can be attractive.

The original application of slipform was on the construction of grain storage silos in North America. Many silos of all sorts have been constructed, and any structure of some height and constant cross-section is a possible subject for this construction method. The original structures all had plain walls, but techniques for constructing doorways, windows and other openings followed, and in consequence various other types of building were constructed, for example multi-storey blocks of flats. Another use is chimneys, while tall bridge piers can often be constructed in this way. While it is simpler to construct only structures with a constant cross-section, it is possible to make changes in layout, either in positive steps, or alternatively continuously. In the former case a structure may have a large plan area for half its height, and the remainder be smaller, or alternatively it may be necessary to reduce a wall thickness. The use of a slipform with a continuous taper has been applied to tall chimneys and to radio towers. As the requirements become more complex, economics dictate that the structures will be higher before it becomes a cheaper method of construction overall.

17.2 DETAILED DESCRIPTION

The actual form panels are not dissimilar from those adopted for traditional formwork. It used to be considered essential to use rift sawn timber boards, but more recently much lower quality timber has been used, as well as various types of plywood and steel. The formwork panels are not connected by form ties, but are held in position by yokes. Each yoke has two vertical legs, one outside each panel, and there is cross bracing above the panels connecting them together. A jack is fixed to the cross-member and this jack takes its support from a rod passing through the centre, whose bottom is at the foot of the wall being constructed. Concrete is placed in this arrangement of form panels at a slow but steady rate, and the jack is used to raise the formwork in small steps to give the same rate of climb.

As well as the basic components just described, the form arrangement includes several working platforms. In plan, a jack and yoke will normally be positioned every 2 to 3 m along the wall, and supported from them will be two platforms about $1\frac{1}{2}$ m below the bottom of the form. The purpose of these platforms is firstly to ensure that the concrete as it is exposed can be rubbed up if it is in any way inadequate, and secondly to enable the supervisor to establish the precise state of the concrete as it emerges. From this the rate of concreting and climbing of the form can be decided. At or near the top of the form will be a platform on at least one side of the wall. This platform will enable concrete and reinforcement to be placed as the form rises. In many cases there will be more platforms, and this depends on the arrangement of the structure and the complexity of the reinforcing steel to be fixed.

17.3 STRUCTURAL DESIGN CONSIDERATIONS

It is pointless to attempt to slide a structure whose design engineer is not committed to sliding as the appropriate method of construction. His consideration must start with the proposal of a layout of the greatest simplicity, compatible with the purpose of the structure, preferably with no sharp internal corners, but instead, splays. The detailing and layout of the reinforcement can present problems if not designed appropriately. Vertical reinforcement should be arranged to lap at various levels, so that at any time during the upward progress of the form there will be some work for the steelfixer to do in lengthening the vertical steel. Horizontal steel should be kept to a minimum, in reasonably short lengths, because it may be necessary to thread it from the outside, and it should as far as possible be devoid of hooks and bends. While it is desirable that the wall thickness and layout remain constant from bottom to top, it is of course sensible to modify the arrangement of reinforcement as the stresses become less.

17.4 EQUIPMENT

The forms themselves are seldom built *in situ* but are prefabricated or may be proprietary steel panels. These are set up on the level base which is necessary for this purpose at the start of the slide. Yokes, obtainable from various proprietary sources, or perhaps purpose made, are set over them, and the deck for distributing steel and concrete is added. It is necessary to have a framework to ensure that the shape of the structure as a whole does not distort as it rises, and this framework is often used as the support for the deck—which will serve as a form for concreting a roof to the structure.

Originally the motive power for slipform was a screwjack manually operated. Nowadays, the normal system is hydraulic. Each jack is connected to the central pump unit, which is operated to drive the jacks up by a pre-determined step between 12 and 40 mm. The frequency of these steps determines the rate of climb. Also used as motive power is air, and electricity. While electricity is of course often used to drive a hydraulic pump, its application can be direct through a motor driving a crank, or even by ribbed wheels acting directly on the climbing rod.

The layout of the form is controlled by the proposed structure. Decisions must be reached on the method of raising and distributing concrete and the means by which the operatives can get up and down. A crane provides a very useful tool for concreting, but standby equipment which can operate in the case of high winds is virtually essential. For operatives, a ladder tower is very useful, but to construct one in advance of the slide presents a fairly major problem, and in consequence other methods are often used. For example, the modern type of rack-operated hoist can be raised as fast as the structure, without impeding its use for access or concreting. Until the advent of ready-mixed concrete there was no alternative but to make concrete on site, but now there is the possibility of sharing this problem with another organisation. If it is carried out on site considerable areas will be needed for a main mixer and a standby in case of mishap to the main mixer. Associated aggregate storage areas will also be needed. In addition, it is desirable to have enough cement storage for the entire operation, to ensure that it is not 'hot'. There must be space on site to lay out all reinforcement in the order required and, where appropriate, door formers and the like for insertion in the concrete.

17.5 ORGANISATION

The operation of a form of this type is normally carried out round the clock. It is possible to work with three 8-hour shifts, but much more popular is the arrangement of two 12-hour shifts. This means that the site manager must provide at least two complete teams. These teams will include a shift supervisor, appropriate trades foremen, access to plant mechanics and electricians, if these are not full time on site; and operatives, primarily labourers and steelfixers, but also joiners where fixings are required. Often joiners can be pursuaded to undertake tasks not normally considered their duty during a slide. The planning of the organisation must ensure that the rate of climb chosen, concrete setting time and concrete facilities proposed are all compatible. While it is always tempting to go faster, there will be a limit due to one of the items listed and in addition really rapid construction may cause unacceptable cracking. Thus a limit to the rate of climb should be determined early in the planning stage. Rates of climb vary considerably, between different organisations, and depend on the particular problems of the structure concerned. It is not practical to change the rate rapidly or frequently, and the ideal is to choose a rate which can be held to over a long period. It is essential that this rate is not so rapid that the concrete is too green when the form exposes it and it falls out, or alternatively that at this time the concrete is so hard that friction between form and concrete causes an overload for the jacks.

17.6 THE START

When the form has been completely assembled, all the site facilities arranged, the gang organised, the pre-determined starting date should have been reached. Concrete is placed in the form in small amounts, of the order of 150–200 mm depth at a time. This is continued so that the overall rate of filling is equivalent to the anticipated rate of climb. After about an hour, an initial lift should be taken. Despite the small age of the concrete, it does not run out the bottom, but this prevents an initial set taking place against the form. When the form is more than half full, a continuous lifting operation is started. By feeling the concrete as it emerges from the form and having a picture of the expected rate of climb, the man directly in charge will judge from the information available to him, just how full to have the form filled and at what rate to have it lifted.

Unless the appearance is of no importance, for example in a lift shaft, it is usual to do some rubbing up. In the case of the ideal concrete mix, there will be sufficient mortar on the surface for this to be rubbed over with a piece of expanded polystyrene to produce a uniform surface very similar to a render coat. Where the mix is harsher, it may be necessary to add some material to achieve this. If difficulties have arisen through faulty compaction, it is normally possible to cure them at this stage.

If for one reason or another some concrete has dropped out of the wall immediately below the form, it is a fairly straightforward matter to deal with. Form panels are made up beforehand in anticipation of this difficulty (it will be too late to start making them when the trouble is found) and these are pressed against the wall obtaining purchase from the hanging scaffold. Concrete is then filled again, and this provides a satisfactory repair.

17.7 HOLES AND POCKETS

For a doorway, the former used is very similar to that used in traditional construction, but its thickness is normally about 3 mm less than the nominal thickness of the wall. This prevents it accidentally being gripped by the shutter and pulled up. Nonetheless it is desirable to anchor such a former with care, to ensure that it is very close to its desired position. Often a run of such opening formers, one above the other, will be connected by bolts so that the lower is always anchoring one above it. Where expanded polystyrene is used as a hole former, it is also cut undersize, to allow a little mortar between it and the form to provide some lubrication to prevent it sticking and being dragged up.

17.8 SPECIAL ASPECTS

Construction of forms for tapered buildings and tapered walls is more complex, but follows basically the same pattern.

The description above refers to slipform which goes up vertically. However, the technique is also applied to the construction of kerbs. It is also practical to use it as a top form to slopes. In this case there is only one form, and it is maintained in its correct position by kentledge.

17.9 CONCLUSION

This method in use for over half a century is a speedy and low cost technique. It may be economic for structures as low as 10 m high, particularly where there are several of them to be constructed one after the other. However, in the normal way a somewhat higher building would prove to be the break-even height. Construction may be by a contractor's own resources, or specialist sub-contractors may be brought in to provide equipment and know how.

Fig. 17.1. A 63 m high slipformed chimney constructed in ten days by two 12-hour shifts. The concrete tapers from 300 to 175 mm over the first 21 m.
(Acknowledgement: Holst & Co. (Northern).)

FIG. 17.2. Slipformed sugar silos and tower.
(Acknowledgement: John Laing Construction Company.)

Fig. 17.3. Slipformed columns to a motorway bridge.
(Acknowledgement: Laing, Iberia.)

Fig. 17.4. Slipformed grain silos in Port of Leith.
(Acknowledgement: John Laing Construction Company.)

Chapter 18

FORM CONSTRUCTION—
TIE ARRANGEMENTS

C. J. WELLER
The Sunderland Forge & Engineering Co. Ltd

The tying of formwork has progressed from the haphazard use of soft iron wire, passed through a structure, windlassed and wedged on to the outer form members, to the closely calculated application of a range of modern 'lost' ties and through bolts. While this progression has obviously been economically desirable, it has also been necessary for tie systems to keep pace with the greater demand imposed by concrete in modern usage as a structural and architectural medium. As technology has advanced, with resulting refinements and complexities, so have the costs increased of formwork that ties support. In this context, form ties, when properly selected, give better value for money than any other piece of equipment on site. This fact becomes very obvious when, for one reason or another, it is necessary to work without them or to use a reduced number of ties.

The first requirement of a tie assembly is that it should safely carry the anticipated load imposed by the fluid concrete, and then possibly fulfil some or all of the following:

(a) Have suitably re-usable components
(b) Leave small or neat holes on the concrete face
(c) Be capable of being 'lost' in architectural features
(d) Be easy to install through heavy reinforcement or large complicated forms
(e) Leave a tie in the construction which can subsequently be used for re-fixing the forms or ancillary equipment

(f) To sustain a load while only stretching within specified limits
(g) Have bearing faces which allow horizontal tying through single or double battered walls
(h) Be capable of single-sided use when necessary

One-piece bolts which pass right through the construction are still used on some general building work, e.g. in moulds for precasting concrete manufacture. Where bolts which actually pass through concrete are concerned this would appear to be the cheapest method of tying; the cost of cardboard or plastic tubes used as sleeves must be taken into account. The more lengthy stripping times involved when withdrawing one continuous bolt right through a completed structure detract from their economy, often causing damage to threads that must be cleaned and re-dyed before their next use. Simple through bolts leave a hole that may require complete filling, no fitting to which the forms can be re-fixed, and once stripped both forms are 'loose' whereas in many cases it may be desirable to keep one face securely tied while the other is stripped.

As an alternative to threaded through bolts, there exists a range of fittings which by virtue of spring loaded cams or wedge actions allow unthreaded bars to be used, the concrete pressure tightening the fittings on to the bar. With most of these fittings special tools are required to tighten the fittings on the bars and to achieve a degree of rigidity in the formwork. Similar special tools are also used to release the re-usable fittings from the bar and to allow their removal. Many situations in form design arise which can only be dealt with economically by the use of through bolts.

Typical of such a situation was the construction of a vertical service shaft, only 1350 mm square in plan running the full height of a multi-storey industrial building. The contractor had made a special collapsing form for the core of the shaft, to be placed at full storey height, which left virtually no room for the operatives to gain access and operate any tie arrangement. In this instance the tie arrangement was restored by welding tapped fittings to the core forms and feeding a threaded bolt through the external forms, the wall void, and into the captive fittings on the core forms.

Similar systems of through bolting are employed on concrete houses and flats, construction allowing in this case all the operations to be performed from inside the house, forms where full periphery

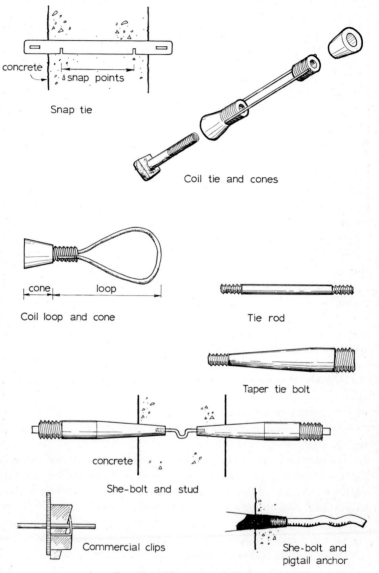

Fig. 18.1. Tie arrangements I.

forms are set up at each floor level. The external fittings are not rigidly held on these forms but retained in a nominal position by a spring-loaded plate which allows the through bolts to 'find' the nut and then pull it normal to the bolt's centre line when the forms are being erected slightly out of line.

A type of tie that combines the principles of through-bolt application and lost-tie arrangements is the snap tie. Snap ties are of round or rectangular steel section and consist essentially of one length of bar with built-in deformations which allow the extreme ends to be snapped off behind the concrete face after the forms are struck leaving only a small blemish. The rectangular variety are used with proprietary steel-framed formwork panels, which fit in recesses on panel to panel joints. In this case the forms are man handled individually as the wedges that secure the ties to the panels also secure a panel to its neighbour. Special deep break back ties are used where groups of sheathing panels are handled by crane.

The round section snap tie, which is far more popular in the USA than in the UK, can be used with all traditional timber forms. Because of their low load-carrying capacity, approximately 1000 kg, very many are required involving a high labour content on any except the very small formwork items. The extreme ends, which are subsequently broken from the face of construction, are made to accommodate a limited range of formwork members, so that standardisation is necessary to enable their use as adaptable, 'go anywhere' fittings.

Until recently the most common 'lost' tie system was generally referred to as the cone and coil type. Here the tie consisted of two or more lengths of mild steel bar welded symmetrically at the ends around a nut formed from a coil of a similar bar. Into these coils a hexagonal or square headed bolt with a matching thread form was screwed to complete the tie arrangement. The bolts passed through the formwork members, into wooden or plastics spacer cones trapped between the ends of the tie and the form face, so the tie and cones together ensured the required wall thickness.

Current varieties of this type possess spacer cones which screw on to the lost ties, some of which are available in high tensile steel which give suitable loadings of up to 7500 kg. A complete assembly of this type would consist of two re-usable bolts, two plate washers, two spacer cones and one lost tie. These coil tie arrangements allow for subsequent fixing for succeeding forms and operations.

An alternative to the cone and coil system is the 'she-bolt' type which consists of internally threaded re-usable bolts that accept round bar lost ties, and upon which a variety of clamps can be screwed to transmit the load from the concrete on to the tie assemblies. Ties with a considerable range of working loads are available, one size being capable of supporting 28 000 kg. The installation of 'she-bolt' tie arrangements into the formwork is as simple as it is for a through bolt, since one does not have to contend with spacer cones. The cost of lost ties is considerably less than the cone and coil types, although the re-usable components tend to be more

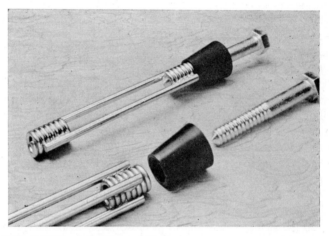

FIG. 18.2. Coil ties and cones provide a means of both tying and spacing the formwork. The buried coil provides anchorage for succeeding operations.

expensive. Certain ties commend themselves to special uses and careful consideration should be made in their selection. At some stage most form tie systems are required to tie back forms in single-sided construction, either as cantilever-type formwork or formwork cast against existing structures, sheet piling and such like. If this work is only part of the construction, then a system of ties and anchors which can be used on all sections of the work provides certain economics based on the interchangeability of components, and simplifies ordering and site stocking.

Once the system to be used has been determined, and taking into account any of the 'use peculiarities' already mentioned and the re-use factor if any, the load which is to be carried should then be calculated. After the pressure has been established, the proportion of the total load on any tie can be apportioned and a factor of safety applied to allow for such matters as a fast rate of placing and possible

FIG. 18.3. The number of through-ties is reduced where strongbacks are used. The soldier members provide access as well as being extensible to produce continuity of support over varying depths of placing.

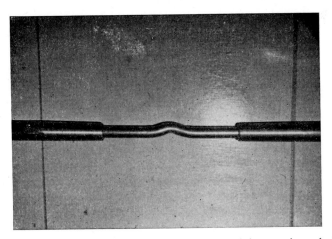

FIG. 18.4. A she-bolt arrangement which shows the stud that remains embedded in the concrete.

prestressing of the ties on assembly. When bolting to an embedded anchor or tie, there is a tendency to overtighten the nuts, a prestressing load becoming superimposed on that resulting from the concrete being cast.

It is imperative that special attention be given to the selection of anchors for single-sided work particularly where climbing cantilever forms are involved. Apart from the *in situ* concrete load, the self-weight of the forms is usually taken by the re-usable bolts that connect on to the anchor; this results in a certain amount of bending in the bolts. Most anchors require a minimum concrete strength of 14 N/mm^2 in which to develop their full working capacity, and if this strength is not reached at the time of loading, the concrete can fail around the anchor. For placings that need to follow on in quick succession, an anchor known as a pigtail can be used since this can develop its load-carrying capacity in concretes of strength as low as 1 N/mm, the length of the anchor being selected to suit the expected concrete strength at the time of loading. Experience has shown that as much as 50% of the expected load can be applied to the anchor by the operatives when clamping the forms in position, and this must be allowed for when selecting the anchor.

Single-sided work against existing structures can cause some problems in achieving support, unfortunately instances do occur where attempts are made to obtain support by using raking props which very often serve to lift the forms as the concrete pressure is transmitted into them. It is often assumed that because there is only one formwork face, the pressure on it must be double that for a double-sided wall, consequently many assume that the pressure is only half that on normal double-sided work and as a result placing can become unsuccessful due to inadequate anchorages being provided.

The most successful, and eventually the cheapest method of tying this type of formwork is to obtain definite fixings into or on to the existing structure. If this is not permitted then the alternatives are: small height castings cantilevered off the previous lift to final height, full height purpose made frames bolted to the floor slab and possibly tied over the waling, or if there is a parallel structure within which *horizontal* props can be used. Where fixings to the existing structure are permitted, almost any of the proprietary expanding anchors can be drilled into position and connected to the preferred tie assemblies. If the existing work is sound concrete or masonry work, no problems will arise, but even so a few random 'pull out' tests should be carried out to check the 'holding' capacity of the old material.

Fixing into old brickwork is always a very dubious practice as brick presents very little resistance to an expansion-type anchor. If

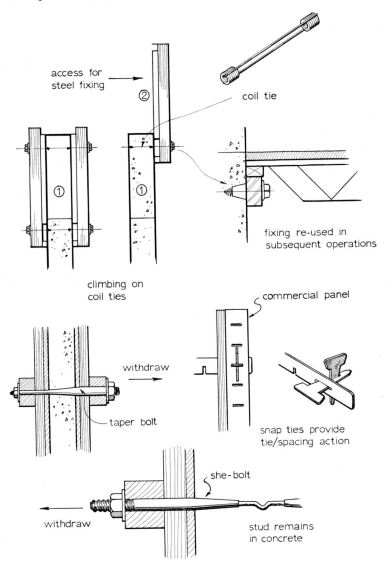

FIG. 18.5. Tie arrangements II.

the depth permits, a threaded bar can be grouted (using cement or epoxy) into the brickwork for subsequent connection to tie rods and these must certainly be tested to arrive at a safe usable load.

Fixings to sheet piling can be much more positive than those adopted for masonry. A nut or coil welded to the piles into which a tie assembly can connect, or a loose nut behind a pair of RS angles (or similar), which are themselves welded across the valley of the piles, gives a certain horizontal tolerance for tie positioning. Nuts positioned behind keyhole slots have been used, but this method cannot be used if the piling is retaining a water-saturated material. If the sheet piling is to be removed after concreting, then ties between it and the new concrete would be a definite embarassment, but here a completely removable 'through' bolt could be used into a nut on the piles, the nut causing little or no resistance to pile removal.

Shot-fired studs have been used for sheet piling for an initial fixing, but it has been found that they do not always generate sufficient holding power to match their full tensile capacity.

The so-called 'waterproof' tie causes more than enough trouble. 'Water bars', 'puddle flanges', 'seals', 'water plates' and a host of other misleading terms are applied to an obstruction on the 'lost' tie, which are believed to be capable of stopping all water penetrating through the wall. These so-called 'barriers' do assist in this respect but under test conditions they have been found to be only 60% effective. If a structure has to be waterproof the extra cost of supplying these 'barriers' makes a nonsense of at least 40% failure rate. The concrete leaks because a horizontal bar in the wall causes an upset in the settlement due to compaction which in turn causes a leakage path to be formed underneath it, and radial cracks around it (see *The Structural Engineer*, April and September 1966, P. Jackson); a barrier must be of considerable size to cover all these potential leakage paths. It seems more reasonable to stop the water penetrating these leakage paths rather than to try and stop it half way along the tie. There are many products and methods that can be adopted to seal the bolt/cone hole on the *water* side of the structure. The most basic materials are a very dry cement mortar, a non-setting mastic covered by a cork proprietary sealing compound or a modern epoxy-type chemical mortar which possess good adhesive qualities and very low shrinkage rate; all these materials stop the water getting to the tie.

The type of tie to be used at any time subject to working load and application will of course depend on a firm's policy, site personnel preference, or cost of the contract. The following exercise gives an indication of the cost differences between the cone/coil system and the she-bolt system of similar load-carrying capacity, based on purchase price, but taking no account of the savings in labour costs between the two. The components are suitable for supporting 3·3 m depth of formwork to a 300 mm thick wall, and the costs are based on the use of 100 no. 8 × 225 mm she-bolt ties, with sufficient re-usable items @ 100 ties at a time. No account is taken for site losses.

Cone and coil equipment SWL 2 000 kg

200 18 × 350 mm bolts	10·20 ⎫
200 PL washers	2·80 ⎬ Re-usable material
200 40 mm spacer cones	0·75 ⎭
100 18 × 220 mm ties	2·50 Lost ties

Total materials for one use   16·25

She-bolt equipment max. ALL 3 500 kg

⎰ 200 14 × 450 mm she-bolts	14·4 ⎱ Re-usable
⎱ 200 clamps	8·2 ⎰ material
100 9 × 225 mm tie rods	1·0 Lost ties

Total materials for one use   23·6

It can be seen from the above that the cost of cone and coil type materials for one use of *re-usable* equipment is only two thirds that of the she-bolt type. For further uses only the lost ties have to be paid for, and with ensuing uses the pattern of cost differences soon becomes reversed as follows:

Cone and Coil		She-bolt	
First use as above	16·25	First use as above	23·6
Sixth use with extra 500 ties @ 12·50	28·75	Sixth use with extra 500 ties @ 5·0	28·6
Tenth use with extra 400 ties @ 10·00	38·75	Tenth use with extra 400 ties @ 4·0	32·6
Twentieth use with extra 1000 ties @ 25·00	63·75	Twentieth use with extra 1000 ties @ 10·0	42·6

It can be seen that at six uses the costs are almost equal, after that every use produces a saving in favour of the she-bolt material and at 20 uses she-bolts have cost only two thirds that of cone and coil type. There are of course many other factors that affect the costs, such as group discounts from one supplier, existing material that can be transferred from another contract, the rate and cost of losses (always very high on work over water), and labour costs which will vary with the system selected and the particular work to which it is applied.

When tie equipment is applied to overhead work in the form of hangers over structural steel, or vertical suspension units to support from above, considerable thought should be given to the safety aspect. If a tie fails, for whatever reason, on a vertical wall placing, the results can be a local bulge in the completed structure (which may or may not be acceptable). It may lead to other ties failing and cause a complete formwork collapse which is very expensive on remedial and replacement costs, but it rarely leads to the death or severe injury of an operative. If a tensile unit fails in overhead work the result will almost certainly lead to personal injury and often to death of operatives.

Chapter 19

THE ACTIVITIES OF THE FORMWORK DESIGN DEPARTMENT

K. ADAMS
Chairman, Joint Formwork Committee, Institution of Structural Engineers

Expressed in simple terms the function of a formwork department is to produce formwork schemes which are economically, as well as technically, satisfactory.

Such a scheme must pass through a series of stages, the pattern of which is similar irrespective of the type of job under consideration. Ideally, this process begins at the tender stage where perusal of the drawings and specifications may result in anything between a brief discussion with the estimator, if the work is of a standard nature, to the preparation of a preliminary design drawing, for pricing purposes, if the work is complex.

Once the job is secured the contract drawings and specifications are then given a very detailed examination. Discussions are held with the planning engineers and site supervisory staff and preliminary drawings are prepared, priced and considered. Very often there are a number of alternative ways of carrying out a job and the merits of each will need to be assessed before a decision is reached. This procedure establishes the approval of a scheme which is acceptable to everyone.

Close liaison is maintained between the site and the design office, if separate from site, during the fabrication, erection, loading and removal of the formwork in order not only to ensure that the work is performed according to the agreed design, but also to monitor the work for any circumstances which may not have been allowed for and to obtain the feedback of information vital to any temporary works department.

The design office also performs a valuable service to the contracting company by maintaining an up-to-date assessment of the cost and performance of the many and varied systems and components on the market. A watchful eye is also kept on the possible potential of new materials, systems and techniques, and it is an unusual department indeed that does not have a development of its own 'on the board'.

19.1 FACTORS THAT DICTATE THE DECISIONS MADE

The factors which dictate a method of construction are many and varied, their relative importance being determined by a host of conditions. The following therefore can only serve to indicate some of the principal considerations. The order does not relate to the degree of importance.

1. Type of finish required.
2. Quality and location of formwork—this will influence the type and style of formwork adopted.
3. Method and rate of concrete placement.
4. Restrictions in lift heights, lengths and areas.
5. Striking times.
6. Programme time.
7. Complexity of structural shape, dimension and pattern.
8. Method of handling formwork, i.e. crane or gantry.
9. Are through ties allowed? Very important if tall wall lifts are planned.
10. Does structure allow the use of a system, i.e. proprietary or possibly the contractor's own?
11. Position of water table in relation to the structure.

Equipment

In common with the permanent works designer the formwork engineer must consider the structural, physical and economic demands of the contract. In addition to this he must also bear in mind the practical aspects of the work. It may be that the problems do not so much lie with the structure to be formed and supported as with the choice between a host of apparently similar items of proprietary equipment.

On occasions the contractor may have in stock equipment suitable for the work, or possibly which, with some adaptation, could be made suitable. On such occasions it is essential to ensure that the availability does not overcome practical considerations. Re-use potential and adaptability, for use on other structures, must also be considered.

At first glance most of the systems and components may appear to have similar characteristics and in turn offer similar advantages. In a competitive market the costs are invariably similar also. In some instances however subtle differences exist which do not manifest themselves until the equipment is in use. This is not to suggest that one system or component is inferior to the next, merely that no one system or component could hope to function equally as well on every type of structure. It might be for example that a system which has performed admirably on one type of structure may not work so well on a similar building, possibly for no other reason than that there is a difference in the centres of the columns and beams or that the arrangement of cross walls is not quite the same.

The construction industry is well served by the specialist proprietary and equipment suppliers, most of whom produce excellent brochures giving details of their equipment together with methods of application. These should be studied at length and comparisons made of their costs, characteristics and performance.

The full potential of a system can only be really appreciated however by the observance of its performance on site. Only by so doing will the designer and draughtsman be able to assess its action and re-action, its weaknesses and peculiarities most of which only become apparent when the system is subjected to any one of a number of varied conditions. Such study forms a further area of the responsibility of the form design department.

19.2 THE IMPORTANCE OF ATTENTION TO DETAIL

The success or failure of a design entirely depends on the approach to detail. It is therefore essential that the importance of this is fully appreciated by the formwork engineer.

In preparing a scheme no assumptions should be made and nothing should be taken for granted. A simple rule is to require proof of every aspect.

For example, is all the material available new or is a percentage secondhand? If secondhand where else has it been used, for what purpose, how many times and under what circumstances? What is its present condition?

A structure is as sound as the foundation upon which it stands and the foundation of good design is based on attentions to detail; this detail must be prepared by the formwork design department.

19.3 PREPARATION OF DRAWINGS

There is no doubt that more attention is paid to work which is properly detailed and hence a higher standard of workmanship is attained. There are a number of aspects which should have special attention. These are:

1. *Detail*. Each drawing should contain clear and concise instructions. While details should not be 'fussy' they should nevertheless leave little to the imagination.
2. *Presentation*. The experienced formwork designer's maxim is 'keep it simple'. If the nature of a structure is complex, every effort should be made to produce a drawing that requires a minimum of explanation on site. The operative will quickly lose sympathy with the designer whose drawing is difficult to follow. Two simple drawings are preferable to one that is complex and the forms detailed invariably cost less to produce. Whereas the designer may have spent a month preparing the drawing, the operative will no doubt be expected to familiarise himself with the scheme within a matter of hours. A good scale, simple concise instructions and clarity of detail are therefore of paramount importance.
3. *References*. Reference to details on other drawings should, where possible, be kept to an absolute minimum and an attempt made to make each drawing as complete within itself as possible. Failure to observe this may make the design difficult to follow and possibly give rise to errors in interpretation.

Extra hours spent on the drawing board may save days on site. They may also ensure a safer structure.

In a typical structure the designer is presented with a range of options. These options will be considered by those employed in the formwork design department and discussed with the site personnel responsible for carrying out the formwork erection on site.

19.4 FLOOR SLAB CONSTRUCTION

With regard to a floor slab construction the options include the use of the following equipment:

Standard adjustable props
Tube and fittings
A proprietary unit support system.

In the case of for example a soffit of 4·5 m height from a suitable support, it is likely that the proprietary unit support system would be selected as being easier to erect and at the same time affording some access during the erection and removal stages.

Where there are substantial conventional downstand beams, these details would point to box beam arrangements. Where the beam configuration divides the slab soffit into a number of rectangular areas this may well tend to lead the designer to use telescopic centres seated on suitably designed beam side numbers.

Such an arrangement can reduce to a minimum the amount of support equipment required. Although, for instance, a bird-cage scaffold may render access easier, factors such as the stocks of system-support equipment in a contractor's yard will be bound to govern the eventual selection of a system.

In the drawings of the beam and slab arrangements produced at this stage it is essential that each item or panel is allocated a mark number. This makes for ease of identification of fabricated panels and ensures that each panel can be relocated in its relative position on subsequent floors.

Each unit of support equipment should also be carefully marked up on the drawings and schedules, and this simplifies taking-off of quantities as well as enabling site personnel to identify every component. All junctions and connections should be detailed and the position and direction of each member indicated. Struts, ties and wedges and plates must be shown, as should the location of such items as grout seals and rule battens at joints.

It is helpful if the designer annotates the drawing with general notes for the user who is concerned with the erection and striking processes, and gives information on good site practice. It is essential that the basic design criteria are stated and such details as the rate of fill and similar factors are set out clearly in the drawing. Where special finishes are concerned it may be necessary to include notes on matters such as form coatings and parting agents.

Walls
If the concrete surface finish is to be of a decorative nature very careful consideration must be given to the detail. This applies not only to patterned surfaces, which may be formed from a rough sawn board or striated or ribbed features but also to the smooth superfine surfaces.

Rationalisation is necessary where the architectural details include such items as door or ventilator openings of varying sizes in a succession of otherwise repetitive faces. Where for instance in a contract there are a series of features such as towers or monoliths of similar dimension which become complicated by the intersection of connecting beams or slabs then there must be considerable detailed discussion with the architect and engineering consultants. The outcome of such discussions carried out by the formwork engineer may well be the subsequent submission discussion and re-submission of drawings outlining contractor's detailed proposals of a practical nature designed to provide the aesthetic results in a manner suited to the contractor's preferred formwork methods.

Proposals may well include the standardisation of lift levels or heights and redesign of such details as board markings or striations to allow standardisation of form height. This ensures consistency of marking on the concrete face while also increasing the form re-use potential.

Lift heights are detailed and so regulated as to allow forms from previous operations to be re-used after some adaptation (this adaptation will be provided for in the framing of the panels as initially constructed and thus expensive on-site time for alterations will be reduced).

Door and ventilator openings may be amended slightly and details adjusted to allow striations and features to be formed neatly and consistently while good compaction and the specified finishes are achieved.

The position of panel joints and of infill strips are regulated and construction arranged to mask connections and thus achieve the consistency of finish required by the authorities. In this situation the designer may seek the agreement of minor adjustments of detail to allow the repetitious use of panels from previous lifts or operations.

On very specialised architectural concrete constructions which are typical of today's prestige buildings, there will still be an extremely large number of individual form panels to be constructed, and in the author's experience the number of such panels has frequently been in the three-figure range on contracts having a particular emphasis on architectural aspects. It is helpful where such large numbers of panels exist to mark them according to the elevation on which they are to be used.

At this stage the designer should be considering the relationship between the inner and outer forms and also changes of finish required where such forms are used to provide striated finishes on one lift and possibly plain concrete on succeeding operations.

Further items for discussion include the treatment and inclusion of weather-seal chases of varying depths and the configuration and interaction of intersecting surfaces, ribs, beams and slabs. All such details must be considered and panels detailed accordingly.

Formwork panels should be detailed on form fabrication drawings in a clear and concise way. The size and section of materials, details of connections and methods of jointing are essential items of information, as well as notes regarding surface treatments, form manufacturing tolerances and similar details.

Special panels such as those instanced take a considerable time to design and detail; they are also expensive to fabricate. A trial section is generally advisable and a inspection should be carried out at the place of manufacture to ensure that the panels are correctly constructed, accurately assembled and that they fit together satisfactorily as well as providing a required standard of form finish to impart the specified quality of finish to the face of the concrete which they are to form.

Appendix

GROUP EXERCISES ON FORMWORK TOPICS

The author wishes to express his thanks to Dr. R. P. Andrew, Director of Training, Cement and Concrete Association, Conference and Training Centre, Fulmer Grange, Slough, Bucks, for permission to publish the formwork exercises which comprise this appendix. These exercises have been devised for use in syndicate work during courses in connection with formwork at that centre.

The exercises are generally carried out by groups of participants who are given the brief, details and drawings. The working sessions require about six hours of concentrated work to allow the participants to return with sensible proposals supported by sketches. The group proposals and sketches are presented to the course and subject to comment both from participants and tutors.

The accompanying details can be seen to offer an excellent opportunity for the student to assess the importance of such factors as system design, re-use capability and materials selection against a commercial background. The examples have been devised to include virtually every aspect of formwork, walls, columns, beams, slabs, ramps, sloping surfaces and thus provide an opportunity for the adoption of a variety of formwork types.

Course tutors are advised to obtain certain basic design aids such as prop selection charts, proprietary formwork, catalogues and similar aids to which students can refer when devising their solutions.

The exercises can be moderated by the tutor for use at various levels of appreciation and the third example indicates such an adjustment to enable the exercises to be used in post-graduate training of engineers concerned with concrete topics. In this instance solutions are requested with regard to a particular part of the structure.

FORMWORK EXERCISE I

General description of the contract

The concrete structure comprises a section of elevated RC road on RC columns and beams approached by two ramps of identical construction. These ramps and the main deck provide access for lorries which will discharge loads of material into the flume structure, the materials being conveyed down the flume and washed into a factory at the northern end of the site.

The external wall of the flume area on the eastern perimeter is to be cast with great care, as it faces a nearby housing estate and school. The architect has decided to express board marking on this wall face.

Schedule of surface finishes

Columns, beams, deck soffits	Specification A
Walls to ramp and flumes generally and external face of kerbs	
External wall face to flume facing east	Specification B
Lining concrete to flumes	Smooth from ply face or cross screeded and twice trowelled

Specification of concrete finishes

(A) This finish is obtained by forming the concrete in properly designed formwork of closely jointed material—steel, ply or timber. The faces of the concrete may bear the imprint of panels and joints. Additionally small surface blemishes caused by entrapped air or water may be expected, but the surface should be free from voids, honeycombing or other large blemishes.

(B) This finish is obtained by forming the concrete off a shutter face comprising treated board, special care being taken to preserve tight joints at construction and panel joints. The boards must be securely supported to avoid deflexion in panels and must be cleaned after each use to avoid staining from release agents.

The joints in the form face must be carefully set out to provide uniform patterning, and in view of the special nature of the finish, the contractor is invited to nominate particular jointing details which will assist in provision of first-class finish.

Striking times specified
Formwork may be removed from vertical surfaces of walls 24 hours from completion of casting.

slabs (props left under)	$3\frac{1}{2}$ days
slab propping	7 days
beam sides	24 hours
beam soffits (props left under)	7 days
beam props	14 days

Alternative striking times will need to be negotiated with the resident engineer.

FORMWORK EXERCISE I (METRIC VERSION)

Object
The object of this exercise is to place you, together with the other members of your syndicate, in a position to understand the various factors affecting the provision of a sound system of formwork for a given contract. You are subjected during the course of the exercise to pressure of time, and will need to allocate to individual members of your syndicate various tasks including that of spokesman for the group. You will have to meet the agent and resident engineer, and discuss with them your intended methods, ensuring that their requirements regarding productivity and engineering principles are covered by your method.

Your brief
You have been appointed formwork designer in charge of formwork on a contract 150 miles from your yard and head office.

Your syndicate is required to devise formwork methods to be used on the contract, and to prepare proposals on the following:

(1) What will be your sequence of construction for the reinforced concrete work from pile caps to completion? Enter this on bar chart and be prepared to discuss your reasons for the sequence chosen.

(2) What time do you intend to allocate to each formwork operation? Indicate duration of each operation on bar chart, outline transparencies and suitable pens will be issued for use on overhead projector.

(3) What recommendations will you make regarding construction

joint positions and joint details? Note and be prepared to report on points which you discuss with the agent and engineer.

(4) What labour force will your trade foreman need to provide the progress needed to complete the formwork task in the allotted time? The papers and drawings herewith give various information which you can use in decision making in course of the exercise, namely:

(A) Programme blank produced by your company planning department showing the time during which you have formwork trades, scaffolders and suitable operatives available, together with information on sub-contractor's starting dates, and the time at which the client must have access to the main deck in order to deal with the seasonal demands of his industry.

(B) A schedule of target rates peculiar to XYZ Co. Ltd, which indicate the level of performance which you may expect on site. The target rates are compounded from tradesmen and labourers' hours, and indicate output to be expected from 1 hour of combined time. There are agreements on site regarding the work which will be undertaken by each grade of operative.

(C) A schedule detailing the concrete finishes on the contract.

(D) Copies of notes made by your contract manager and agent at a meeting with the architect and engineer.

Notes from meeting with architect and engineer
The engineer agreed to a joint 50 mm above deck level, to assist in formation of edge or kerb beams (non-structural).

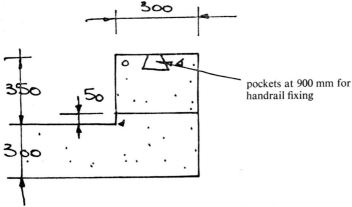

FIG. A.1. Sketch on deck at perimeter.

The architect and engineer each expressed their anxiety to provide straight line joints on the face of the concrete structure.

The architect agreed to tie holes showing on the concrete face, and the engineer requests information on the methods we will adopt for tying wall shutters.

The engineer wishes to hear at an early stage in the planning of the contract where we wish to establish construction joints, and our proposed method of formation of these joints.

The architect expressed concern that his client should have access to the main deck by September, to take full advantage of his investment in the new structure. It was agreed that the client would use the main deck, gaining access via the south ramp only. The engineer agreed that the main deck could be used 28 days from casting last bay of main decking.

Our formwork designer must ensure that he co-operates fully with the site agent in the matter of providing continuity of concrete work, and sufficient formwork to give economical casting operations.

A check on site conditions has been made and the ground is apparently sound, with a low water table—thus we should not have trouble with work in column bases and excavations.

The engineer is particularly insistent that the walling should be cast in bays of not more than 6 m length. Bays should mature for at least 7 days before the adjacent lift of concrete is cast.

Contacts and communication

The design engineer is available to answer your questions and discuss proposals which you may make regarding height of lifts, position of construction joints, sequence of casting etc. He will visit your group during the course of the exercise.

The site agent for XYZ Co. Ltd will also be available to answer questions on concreting capacity, amounts and rate of steel fixing etc. He will be vitally interested in the economics of the methods which you adopt.

Plant and materials

There will be a crane available on the site when required to handle various items of formwork. For guidance assume crane with 24 m jib lifting 7 tonne at 15 m radius. The crane which is available is on tracks. Extra jib sections may be inserted if required. There are reasonable supplies of all materials available when required, e.g. props, scaffold, plates, ply and timber.

Formwork quantities
N.B. Construction joints are not measured; these are included in concreting rates.

South ramp

Strip foundation 450 deep	45 m²
Walling to south ramp 150 thick class A finish, height up to 2250	111 m²
Columns to south ramp 375 sq section heights 1350 to 4800 class A finish	51 m²
Upstand kerb to ramp 350 deep class A finish	70 m²
Main beams to ramp 900 deep × 350 wide height 1350 to 5700	71 m²
Soffit to ramp slab class A finish	172 m²
Edging to ramp 300 deep class A finish	54 m²
Rough shuttering to concrete in ground 750 average deep	46 m²

These quantities repeat in north ramp

Main elevated road deck

Rough shuttering to concrete in ground	124 m²
Column to main deck 600 sq class A finish	214 m²
Beams to main deck 900 deep × 600 wide class A finish	360 m²
Soffit to main deck class A finish	700 m²
Edging to main deck 300 deep class A finish	109 m lin
Upstand kerb 350 deep class A finish	78 m²

Flume walls and flumes

Strip foundation beam 600 deep rough shuttering	131 m²
Walling specification A up to 6600 high	1143 m²
Walling special board marked finish up to 6600 high class B finish	285 m²
Dwarf wall base slab edging 225 wide rough board shuttering	84 m lin
Dwarf walls to flume 150 thickness × 1800 high class A surface finish	311 m²

Provide profiles and formwork to upper surface of infill 150 thick 481 m²

Provide screedboards to *in situ* benching to flume drainage channel 84 m lin

Standard rates for formwork manufacture and erection
Applicable to XYZ Co. Ltd only
Hours quoted are compounded from carpenter and labourer hours

Fabricate beam and column forms in timber to specification A	0·250 m²/hr
Fabricate rough shuttering suitable for ground beams, concrete in ground, etc.	0·750 m²/hr
Erect rough shuttering including all strutting lining and levelling	0·750 m²/hr
Erect column shuttering including strutting up to 3300 from ground level	0·500 m²/hr
Ditto do 3300 to 6000 from ground level	0·420 m²/hr
Erect beam soffits and beam sides up to 3300 from ground level including all bracing close jointed to specification A	0·250 m²/hr
Ditto about 3300 from ground	0·200 m²/hr
Fabricate wall shuttering in panels or units suitably close jointed to cast finished concrete to specification A	0·250 m²/hr
Line existing shutters with ply or board to provide board marked or textured surface	0·750 m²/hr
Erect proprietary decking/soffit shutters to slab including all strutting lining and levelling not exceeding 3300 from ground	1·100 m²/hr
Ditto exceeding 3300 from ground	0·750 m²/hr
Erect wall shuttering including all propping, ties lining and levelling not exceeding 3300 from ground	0·750 m²/hr
Ditto exceeding 3300 from ground	0·500 m²/hr
Erect screeding profiles to sloping surfaces cast upon infill including all strutting	0·250 m²/hr
Erect formwork to upper surface of sloping slab	0·200 m²/hr
Erect edge beam formwork including strutting	0·250 m²/hr

The target rates allow for infill, strutting and all lining and levelling. Allowance of 25% of erection time has been made for striking and carting to next location, including for unskilled assistance in slinging, etc.

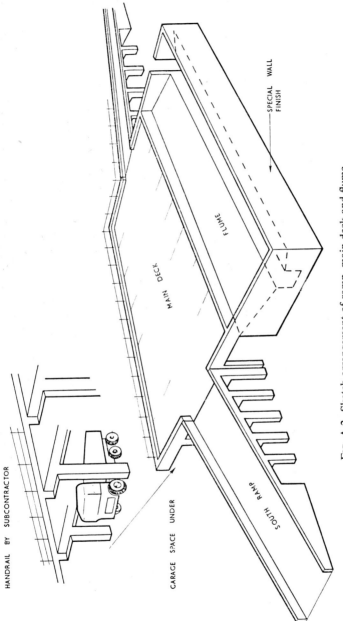

Fig. A.2. Sketch arrangement of ramp, main deck and flume.

270   Practical Formwork and Mould Construction

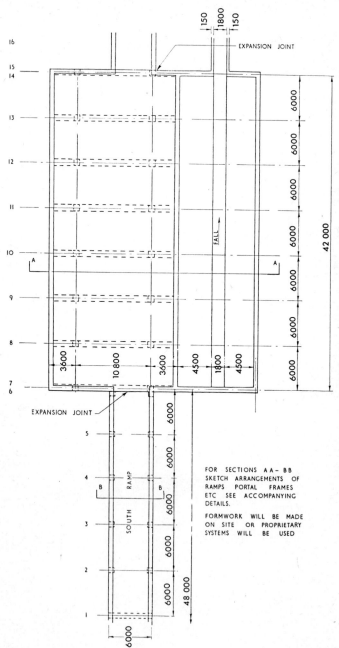

Fig. A.3. Part plan on proposed elevated roadway, garage and flume.

Appendix—Group Exercises on Formwork Topics 271

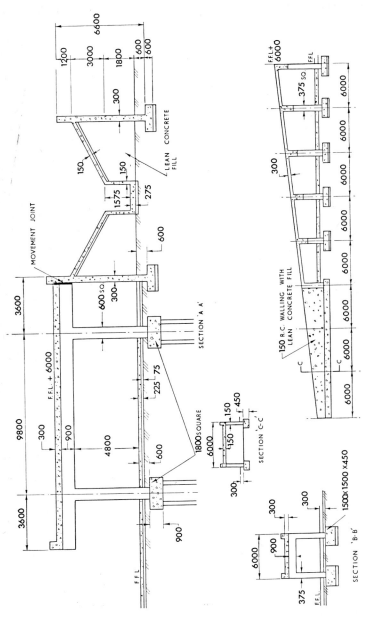

Fig. A.4. Longitudinal section on ramp. *Note*: both ramps are identical.

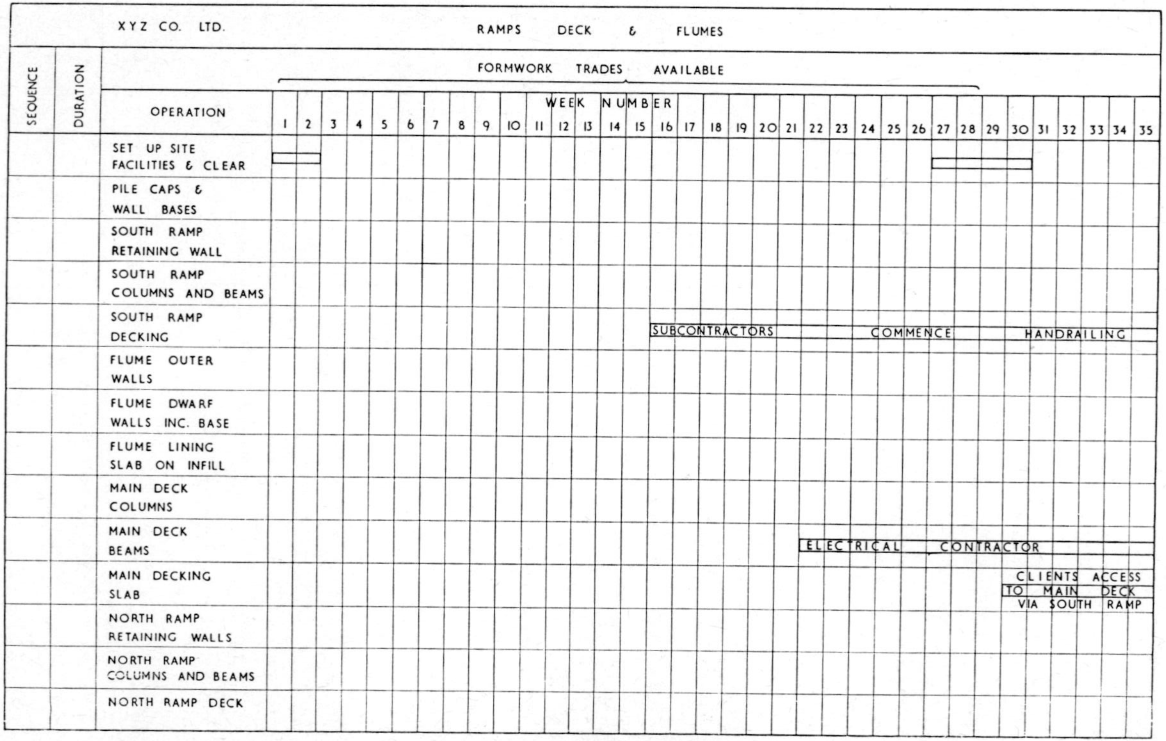
XYZ CO. LTD.
RAMPS DECK & FLUMES
FORMWORK TRADES AVAILABLE
SEQUENCE
DURATION
OPERATION
WEEK NUMBER
1 2 3 4 5 6 7 8 9 10 11 12 13 14 15 16 17 18 19 20 21 22 23 24 25 26 27 28 29 30 31 32 33 34 35
SET UP SITE FACILITIES & CLEAR
PILE CAPS & WALL BASES
SOUTH RAMP RETAINING WALL
SOUTH RAMP COLUMNS AND BEAMS
SOUTH RAMP DECKING
FLUME OUTER WALLS
FLUME DWARF WALLS INC. BASE
FLUME LINING SLAB ON INFILL
MAIN DECK COLUMNS
MAIN DECK BEAMS
MAIN DECKING SLAB
NORTH RAMP RETAINING WALLS
NORTH RAMP COLUMNS AND BEAMS
NORTH RAMP DECK
SUBCONTRACTORS COMMENCE HANDRAILING
ELECTRICAL CONTRACTOR
CLIENTS ACCESS TO MAIN DECK VIA SOUTH RAMP

FIG. A.5. Programme blank.

FORMWORK EXERCISE II

Object

The object of this exercise is to place you together with the other members of your syndicate, in a position to understand the various problems to be encountered by the formwork designer attempting to provide a sound system of formwork for a given contract.

You are subjected during the course of the exercise to the pressures of time, and will need to allocate to individual members of your syndicate various tasks including that of spokesman for the group. You will have to meet the agent and resident engineer, and discuss with them your intended methods, ensuring that their requirements regarding productivity and engineering principles are covered by the methods which you propose to adopt.

Your syndicate

You are appointed formwork designer for the ABX Construction Company based at their head office, with responsibility for liaison with engineers and architects as necessary. You are also expected to maintain liaison with the agent, foreman and trades foreman involved in the structural reinforced concrete work carried out by the company.

You are required to design the formwork methods and systems for the new grain silo which your company are to construct at a site some 150 miles from head office and yard. All excavation work is being carried out by sub-contractors. You are requested to prepare proposals and report on the following at the forthcoming meeting with the various members of your company:

(1) What will be your sequence of construction for the reinforced concrete work from foundation slab and column caps through underbin and bin walls to completion of parapet walls? Complete a bar chart and provide any other planning information which will support your proposals.

(2) What will be the timing which you propose for the individual operations? Indicate the duration of each operation on a bar chart. (Transparencies and suitable pens will be issued for use with the overhead projector.)

(3) What recommendations will you make regarding construction joint positions and joint details? Note and be prepared to report on points which you raise with the engineer and agent

who will be available for discussion during the course of operations. The engineer will also discuss reinforcing steel positions and other details of the structural design which you may wish to raise.
(4) What labour force do you anticipate will be required on site to meet the formwork programme which you set out?
(5) Produce sketches (on transparencies for use with the overhead projector) which indicate the construction of your formwork to columns, beams, walls, floor slabs and hopper bottoms, together with any constructional details which expand your proposals.
(6) Detail your methods of form handling and access arrangements for form erection, steel fixing and concreting.

Notes from meeting with architect and engineer
The architect and the engineer each expressed their desire to provide straight line joints on the face of the structure.

The architect agreed to tie holes, or bolt holes, showing on the concrete face, and the engineer will need to approve the method of tying which we propose to adopt for wall formwork.

The engineer wishes to hear at an early stage in the contract where we wish to establish our construction joints and our proposed method of forming these joints.

The architect was concerned that the plant engineer should have access to the plant tower at the time specified, as the erection of the elevators and chutes forms an essential part of the process plant which the client has to use immediately on occupation.

The engineer requires that particular consideration should be given to provision of standing supports or props to the floors in the plant tower whilst plant erection is in progress; we must submit sketches of our proposals.

A check on site conditions has been made and the ground is apparently sound with a low water table; thus we should not have difficulty with work in excavations and foundations.

The engineer is prepared to consider various ways of constructing the walls in lifts, and will discuss the specification item which requires that concrete should not be placed in bays exceeding 6 m in length, and that at least 7 days should elapse between the casting of adjacent bays.

The engineer has agreed to the use of kickers, but has stipulated that these should not be less than 75 mm high, and that they should be cast monolithic with the previous lift of concrete whenever practical.

Schedule of surface finishes

Columns, beam sides and soffits, slab soffits up to canopy level — Specification A

External wall faces to main silo structure. Gable wall to plant tower — Specification B

Internal faces to bins — Specification C

Specification of concrete finishes

(A) This finish is obtained by forming the concrete in properly designed formwork of closely jointed material—steel, ply or timber. The faces of the concrete may bear the imprint of panels and joints. Additionally, small surface blemishes caused by entrapped air or water may be expected, but the surface should be free from voids, honeycombing, or other large blemishes.

(B) This finish is obtained by forming the concrete off a shutter face comprising plastic faced board or steel panels, special care being taken to preserve tight joints at construction and panel joints. The individual boards must be securely supported to avoid deflection in panels and must be cleaned after each use to avoid staining from release agents if used.

The joints in the form face must be carefully set out to provide uniform patterning, and in view of the special nature of the finish, the contractor is invited to nominate particular jointing details which will assist in provision of first-class finish.

(C) This finish is obtained by casting from panels or boards to provide a perfectly smooth finish, approaching mirror finish, and certainly of a smooth self-cleaning nature to avoid contamination and build-up of grout on the surface. Particular care must be taken to avoid any steps, lips or other obstructions to the free flow of grain down the surface.

Striking times specified

Formwork may be removed from the faces of the concrete in the times specified:

walling	12–14 h by consent
slabs (props left under)	$3\frac{1}{2}$ days

slab propping 7 days
beam sides 24 hours
beam soffits (props left under) 7 days
beam props 14 days

Alternative striking times will need to be negotiated with the resident engineer.

Standard rates for formwork manufacture and erection
Applicable to ABX Construction Company only
Hours quoted are compounded from carpenter and labourer hours

Fabricate beam and column forms in timber to specification A	0·25 m²
Fabricate rough shuttering suitable for ground beams, pile caps etc.	0·75 m²
Erect rough shuttering including all strutting lining and levelling	0·75 m²
Erect column shuttering including strutting	0·5 m²
Erect beam soffits and beam sides including all bracing close jointed to specification A and B	0·25 m²
Fabricate wall shuttering in panels or units suitably close jointed to cast finished concrete to specification A, B and C	0·25 m²
Erect decking/soffit shutters to slabs including all strutting lining and levelling to specification A and B	0·75 m²
Erect wall shuttering including all propping, ties, lining and levelling to specification A	2·5 m²
Specification B and C	1·5 m²
Erect formwork to upper surface of sloping slab	0·25 m²

The target rates allow for infill, strutting and all lining and levelling.

Allowance of 25% of erection time has been made for striking and carting to next locations, including for unskilled assistance in slinging, etc.

General description of the contract
The structure is a reinforced concrete grain silo comprising 12 main bins and 8 small bins in which grain is stored after harvesting, prior to discharge into railway vehicles or road transport, for distribution throughout the country. At the north end of the bin area is a multi-storey plant tower which houses chutes and elevators used in

handling the grain. The floor slabs in this area are supported by columns and beams, with a major movement joint between the tower and the north bin wall. Adjacent to the main silo is a covered area formed by reinforced concrete slab upon portal frames; the slab is pierced by circular rooflights. The whole of the road to the east of the main bins will have a 225 mm slab for vehicular access; the rail side at the west having rail track on sleepers and ballast based upon well compacted hardcore.

To avoid complications at the planning stage, the dimensions of through-holes and openings in the gable wall and floor slabs to the plant tower and overbin slab are omitted.

Reinforcing steel is generally of a light nature in walls and slabs, although in the raft and underbin slabs there will be considerable steel as in the concrete fillets in the main bin corners which form columns within the completed work.

The completion dates for the concrete structure must be such that the client may take up occupation in time to receive grain consignments in early August, by rail, i.e. 12 months from commencement.

Contacts and communication
The resident engineer is available to answer your questions and discuss proposals which you may make regarding the height of lifts, position of construction joints and any other modifications which you wish to make to expedite progress on the job.

The site agent for ABX Construction Company will also be available to answer questions on concrete capacity, rate of steel fixing etc. He will be vitally interested in the economics of the systems which you adopt, and will advise upon the general suitability of the formwork methods which you suggest within his overall plan for carrying out the work.

For the purpose of the exercise you will be well advised to consider the formwork aspects in particular, and accept that the agent will ensure that sub-contractors and other trades meet their commitments as to progress and completion.

Plant and materials
You will be able to take advantage of the fact that your company has at present large holdings of props, scaffold, timber plates and similar equipment, which has recently come available from a large contract

nearing completion. You have a free hand regarding choice of timber, ply, glass reinforced plastics, or steel as form materials, and may, if you wish, nominate the type of proprietary formwork which you propose to use.

The company policy regarding craneage is that they use tower cranes on contracts of this nature. A crane will become available 5 weeks prior to your commencement on site. It will have ample lifting capacity over the complete area of the structure.

Drawings and information

The papers and drawings herewith give various information which will be useful in decision making in the course of planning the contract, namely:

(A) General assembly drawings and sketches of the project, together with sketch arrangement for identification purposes.

(B) A schedule detailing the concrete finishes on the contract.

(C) A schedule of striking times for formwork.

(D) A schedule of target rates peculiar to **ABX** Construction Company, which indicates the level of performance which you may expect on site. The target rates are compounded from tradesmen and labourers' hours, and indicate the output to be expected from one hour of combined time. There are agreements on site regarding the work which will be undertaken by each grade of operative.

(E) Copies of notes made by your contract manager and agent at a meeting with the architect and engineer.

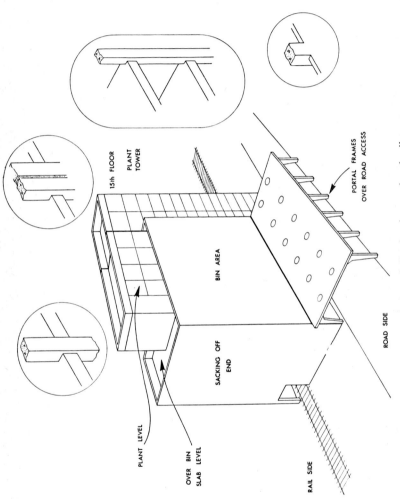

Fig. A.6. General view of silo showing road and rail access.

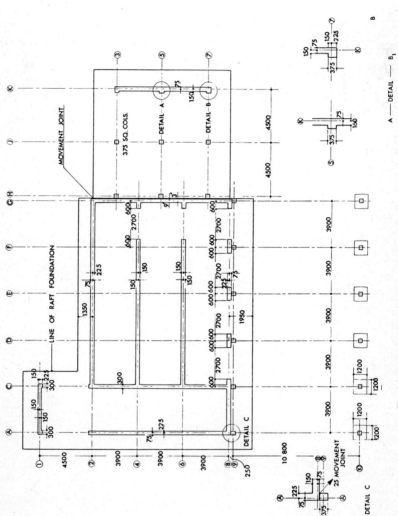

Fig. A.7. Foundation plan showing walls at underbin level.

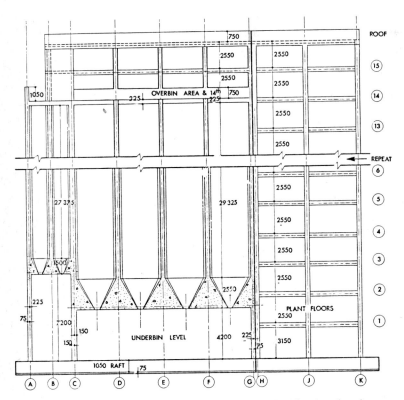

FIG. A.8. Longitudinal section between grid lines 7 and 8 (showing elevation on plant tower and structure above overbin slab).

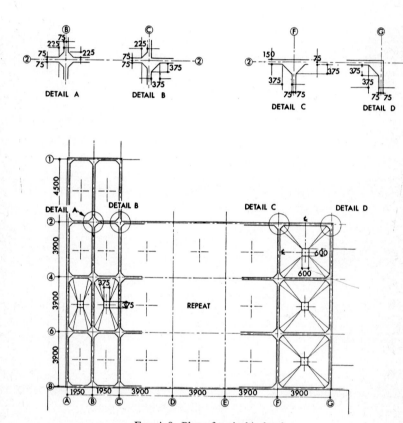

FIG. A.9. Plan of main bin level.

Appendix—Group Exercises on Formwork Topics 283

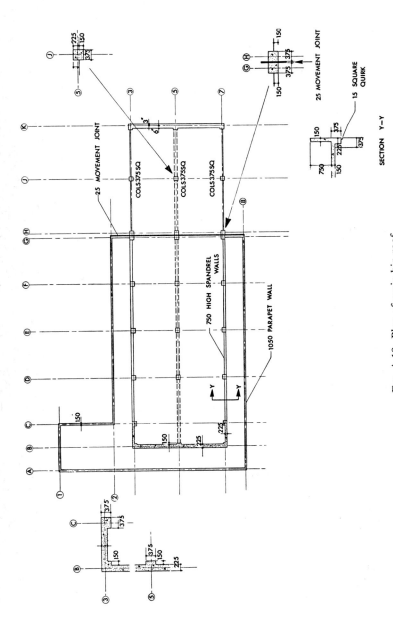

Fig. A.10. Plan of main bin roof.

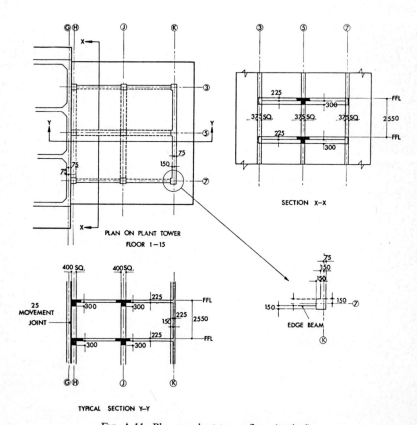

Fig. A.11. Plan on plant tower floor (typical).

Appendix—Group Exercises on Formwork Topics

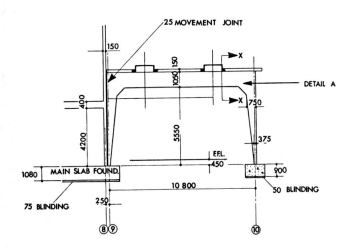

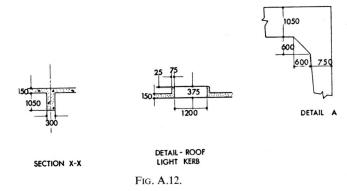

Fig. A.12.

PLANNING A CONCRETING OPERATION

Your brief
This brief is to be read in conjunction with information given in TDH 1602.

You have been requested to assist the site agent responsible for the construction of a grain silo. One section of the work calls for particular attention and the agent has asked for the services of an engineer to plan and control this part of the work.

Area of operations
You are required to deal with *all* operations in casting bin bottom concrete between raft+4·2 m and raft+7.2 m.

Requirements
Working in syndicate you are required to set down recommendations for the following:

1. The position of day joints and construction joints and a suitable sequence of operations.
2. Means of handling, placing and compacting the concrete noting labour and equipment requirements.
3. Establish the necessary labour force to carry out the operations in the *main bins* between levels raft+4·2 m to raft+7·2 m in a period of 8 weeks.
4. Provide outline sketches of main arrangements of formwork, make suggestions as to work which can be carried out in preparation for these operations.

You can obtain answers to any queries arising from the project organiser.

BIBLIOGRAPHY

AKROYD, T. N. W. (1969). The responsibilities for falsework and temporary works. Paper presented at a symposium on Falsework, arranged by the Institution of Structural Engineers and The Concrete Society at the Royal Aeronautical Society, London, 3 February 1969, London, The Concrete Society, p. 7.

AUSTIN, C. K. (1966). *Formwork to Concrete*, London, Cleaver-Hume and Macmillan incorporating Cleaver-Hume. Second edition, pp. 283 and 308.

BIRCH, N. et al. (1971). Effect of site factors on the load capacities of adjustable steel props. London, Construction Industry Research and Information Association. Report No. 27, p. 48.

BRITISH STANDARDS INSTITUTION. Metal scaffolding, BS 1139:1964, London, British Standards Institution, p. 48.

BRITISH STANDARDS INSTITUTION. Steel nails, BS 1202:Part 1:1966; Copper nails, Part 2:1966; Aluminium nails, Part 3:1962, London, British Standards Institution, p. 56.

BRITISH STANDARDS INSTITUTION. Plywood manufactured from tropical hardwoods, BS 1455:1963, London, British Standards Institution, p. 20.

BRITISH STANDARDS INSTITUTION. Connectors for timber, BS 1579:1960, London, British Standards Institution, p. 24.

BRITISH STANDARDS INSTITUTION. Metal props and struts, BS 4074:1966, London, British Standards Institution, p. 12.

BRITISH STANDARDS INSTITUTION. Glossary of formwork terms, BS 4340:1968, London, British Standards Institution, p. 28.

CHAMPION, S. (1969). Falsework with tubes and scaffold fittings, props and proprietary systems. Paper presented at a symposium on Falsework arranged by the Institution of Structural Engineers and The Concrete Society at the Royal Aeronautical Society, London, 3 February 1969, London, The Concrete Society, p. 9. Technical Paper PCS 34.

THE CONCRETE SOCIETY (1971). Falsework. Report of the Joint Committee of The Concrete Society and the Institution of Structural Engineers, London, The Concrete Society, p. 52.

THE CONCRETE SOCIETY (1974). Formwork. Report of the Joint Committee of The Concrete Society and the Institution of Structural Engineers (in draft form).

CONSTRUCTION INDUSTRY RESEARCH AND INFORMATION ASSOCIATION (1969). Formwork loading design sheets: (1) Metric. (2) Imperial. London, CIRIA.

THE COUNCIL FOR CODES OF PRACTICE. The structural use of timber, CP 112:1967; Metric units, CP 112:Part 2:1971, London, British Standards Institution, p. 128.

HURD, M. K. *Formwork for Concrete*. Prepared with the assistance of R. C. Baldwin under the direction of ACI Committee 622. Second edition.

JESSOP, K. G. (December 1970). Steel formwork: accuracy money or myth? *Civil Engineering and Public Works Review*, **65**, No. 773, pp. 1439–1442, 1468.

KINNEAR, R. G. (1964). Concrete surface blemishes. A classification of the surface defects and some particular influences of formwork linings, release agents and concrete pressure on the appearance of concrete finishes. Paper presented to the CIB Working Group W.29—Concrete Surface Finishes at Norwegian Building Research Institute, Oslo, 30 August–1 September 1963. Technical Report TRA 380 (July 1964), London, Cement and Concrete Association, p. 36.

LUNDIN, T. (1970). Former-speciella problem (Concrete forms—special problems relating to moulds for precast concrete elements), *Nordisk Betong.*, **14**, Part 3, pp. 215–240.

RICHARDSON, J. G. (1972). Concrete as a mould material. Paper submitted to the BIBM Congress, Barcelona.

RICHARDSON, J. G. (May 1972). *Formwork Notebook*. London, Cement and Concrete Association, p. 94.

RICHARDSON, J. G. (1973). Practical considerations in the provision of formwork for structural concrete. Paper submitted to the ACI Spring Convention, New Jersey, USA.

RICHARDSON, J. G. (September 1973). *Precast Concrete Production*, London, Cement and Concrete Association, p. 232.

RICHARDSON, J. G. (January 1975). *Concrete Notebook*, London, Cement and Concrete Association, p. 118.

SNOW, F. (1966). *Formwork for Modern Structures*, London, Chapman and Hall.

WEAVER, J. and SADGROVE, B. M. (1971). Striking times of formwork—tables of curing periods to achieve given strengths, CIRIA Report 36.

INDEX

Access
  changes in, 30
  working platforms, 30
Accuracy
  attainment of, 25
  dimensional stability, 26
  establishing critical dimensions, 26
  jigs and templates, 102, 104
  moulds, of, 53
  running dimensions, 65, 70
Activity relationships, 35, 36
Architect
  interest in formwork, 4
  requiring guidance, 6
  using concrete, 5

Backing members, 51
Beams
  anchoring plates for floors, 151
  cast prior to main slab, 150
  circular, 153
  clamps, 152
  erection of forms, 149
  haunched, 150
  panel layouts, 66
  post tensioned, 150
  striking, 28
Bins and silos, handling panels, 67

Calculation for sectional sizes, 64
Cambers in precast beams, 184
Cantilevers, beams, 157
Casing steel, 160

Casting cycle, 15
Centres
  barrels, to, 78
  flooring, in, 77
Circular work
  facets to, 51
  kickers to, 114
  using adjustable forms, 138
  using hardboard sheathing, 90
  using ply sheathing, 89
Column forms
  adjusting height, 146
  bracing to, 116
  circular, 144
  construction, 140
  fabrication, 142
  joints in, 24
  landing rings, 141
  linked columns, 147
  mushroom heads, 145
  sheathing, 140
  size adjustment, 149
  splayed, 143
  striking times, 28
Concrete
  appearance, 9
  joint positions, 43
  pallets, 175
  placing considerations, 22, 49
  slipforming, 239
Construction joints
  consultation, 11
  establishing position, 43
  formation of, 165
  groundwork, in, 112
Continuity of face, 51
Contractor's interest in formwork, 7

Contracts
  interchange of ideas, 14
  progress, 13
Curing
  accelerated, 170
  continuity, 28
  cubes for striking purposes, 28
  non-critical time, in, 111
Cutting lists
  preparation in design department, 13
  timber materials, for, 102

Deflection
  related to support spacing, 21
  resulting from partial fill, 130
  single-sided work, 128
Design
  consultants, by, 11
  contractor's design department, by, 12, 255
  definition, 1
  factors governing decisions, 256
  impact on results, 1
  local arrangements for, 14
  practical nature of, 13
  supervision, 14
Design department
  activities of, 255
  availability as consultants, 14
  details produced, 13
  savings resulting from, 12
Drawings
  general assembly, information on, 61
  general assembly, value of, 57
  limiting numbers of, 65
  precast outlines, value of, 69
  presentation, 258
  scales for, 67
  sectional details, 61
  sketches, value of, 67
  standard procedures, 60
Ducts
  post tensioned concrete, in, 185
  slabs, in, 115

Engineer
  concern with accuracy, 6
  providing drawings, 7
  relationships with form design, 7
  responsibility, 14
Equipment
  economics, 257
  re-use considerations, 14

Features
  architectural, 191
  avoiding feather edges, 86
  change of texture, at, 193
  clearance at demoulding, 32
Fines retention, 20
Fixings
  brick and masonry, for, 146
  nailing, 117
  screws, 86, 119
  staples, 119
  various materials, 252
Flewing surfaces, 90
Floor forms
  bay layouts, 66
  opening formers, 90
  supporting joists, 151
Footings
  struts, to, 109
  supports, to, 22
Formers
  alloy, 95
  concrete, 115
  door openings, 126
  flotation, 96
  hollow, 95
  inflatable, 95
  openings, to, 90
  pockets, to, 115
  precasting, in, 171
  striking, 131
Formwork
  abutments, 121
  beams, 149
  chimneys, 235
  circular and conical, 135
  cleaning, 87, 127
  column bases, 115

Index

Formwork—*contd.*
  columns, 140
  crossheads, 229
  drop panels, 162
  ducts and pockets, 113
  foundations, 108
  framed panels, 159
  handling, 30
  inspection, 10
  large forms, 226
  panel sizes, 30
  silos, 235
  slipform, 235
  tall columns, 228
  waling, 119, 260
Formwork exercises
  application, 262
  elevated roadway, 263
  grain silo, 273
  planning a concreting operation 286
Foundations
  back casting, 110, 111
  column bases, 115
  ground beams, 108

Ganging
  columns, 205
  piles, 170
  purlins, 171
Geometrical work, 103
Grillages
  hollow floors, for, 163
  mould construction, in, 180
Grout loss, 19

Handling
  abnormal movements, 47
  access for cranes, 33
  affecting panel sizes, 32
  angleform, 203
  considerations, 34
  cross wall forms, 132
  discussion over, 37
  minimising operations, 46
  plant limitations, 54

Handling—*contd.*
  tableform, 199
  types of crane, 32
  using chain blocks, 134
  wall panels, 133
Hole formation, 56
Hopper form adjustment, 51

Infill
  construction, 48
  flooring, in, 160
  modular systems, in, 74
  over several lifts, 49

Kickers
  columns, 112
  foundations, 112
  installation, 36
  single-sided, 128
  sloping slabs, 14

Lap, 51
Lead on formers, 63
Long line moulds, 168

Machinery
  metal working, 105
  woodworking, 99
Manufacture
  moulds and forms, 98
  prior to erection, 57
Materials
  alloys, 94
  choice of, 39, 58
  clay, 191
  concrete, 91
  foam, 192
  hardboard, 90
  hardwood, 84
  interaction with construction, 39
  miscellaneous, 94
  particle board, 91
  plaster, 91, 191, 223
  plastics, 92, 193, 208

Materials—*contd.*
  plywood, 87
  rubber, 193
  softwood, 83
  steel, 72, 225
  steel banding, 145
Mould
  assembly, 101
  assembly after stripping, 32
  backing members, 176
  construction, 68, 174, 179
  detail drawings, 68
  early easing, 29
  inspection, 69, 103
  mechanisation, 173
  modifications, 70
  reducing loose parts, 34
  sculptured work, 193
  sheathing, 177

Nibs
  columns, 147
  formation, 50
  precast, 55
  precasting, in, 181
  returns, at, 120

Openings
  lead, 63
  precast beams, in, 181
  slipforming, in, 240
Optimum profile
  columns, of, 148
  establishment of, 43

Panels
  adapting to profile, 48
  detail drawings, 64
  establishing height, 49
  establishing profiles, 47
  numbering on layout, 67
Personnel
  contractor's agent, 8
  mill foreman, 101
  setter-out, 101

Personnel—*contd.*
  site engineer, 8
  trades foreman, 9
Pile casting, 169
Placing technique
  avoidance of bridging, 23
  effect of rate of rise, 22
  reduced height panels, 126
  special formwork, in, 232
Planning
  associated activities, 35, 37
  beam sides and bases, 168
  consultants, by, 11
  drawing availability, 36
  lifts, 38, 46
  slab formwork, 157
  slipforming, 238
  spandrel casting, 132
  value of models, 46
Plastics
  advantages, 208
  foamed, 220
  glass reinforced, 209
  mould material, as, 208
  polyurethanes, 217
  thermoplastics, 215
  vinyl rubber, 219
Plate footings, 109
Plywood
  hardwood faced, 88
  impregnated, 89
  reducing number of joints, 89
  stopends, in, 165
  treatment, 87
Precast details
  aperture formation, 56
  early receipt, 54
  establishing profile, 54
  nib and corbel formation, 55
  pallet design, 55
Precast moulds
  bridge beams, 183
  culverts, 188
  floor units, 171
  frame units, 175
  piles, 169
  products, 189
  purlins, 171

Proprietary formwork
  angleforms, 195
  centres, 160
  crosswall forms, 132
  double-headed props, 156
  floors, 159
  large steel panels, 194
  tableform, 195
  tunnel forms, 202
Props
  adjustment, 76
  double-headed, 28, 156
  easing, 28
  formal layout, 66
  push-pull, 132
  safety, 76

Quality control
  inspection, 10
  materials aspects, 20

Reinforcement
  cranking, 50
  projecting, 183
  required support, 133
  sequence of fixing, 230
  starters, 131
  storage, 23
Release agents, 80
Reports
  Bragg Committee, 232
  Falsework Report, 232
Reshoring, 28
Responsibility, 8
Retarders
  lacquer type, 82
  use of, 81
Re-use
  determination of numbers, 48
  standardisation for, 167
  thicknessed timber, of, 20
  timber forms, of, 85

Safety, 162
Scaffold, 152

Schedules, 60
Shaped ribs
  circular beams, in, 153
  conical work, in, 136
  waling, in, 135
She bolts, 123
Slipform
  construction, 236
  economy, 235
  equipment, 237
Soldiers
  single-sided work, in, 128
  supporting chutes, 126
Spacers, 118
Spiral stairs, 153
Stack casting, 184
Stairs and landings
  corners, 49
  special forms, 204
  starters to, 130
Steel formwork
  obtaining close joints, 20
  proprietary modular systems, 73
  purpose made, 72, 226
Stopends
  calculation for, 166
  construction, 165
  expanded metal, 164
  hardboard, 90
  plumb of, 166
  precast concrete, in, 181
  prestressed, in, 56
  removal, 28
Storage
  between uses, 33
  safety aspects, 33
  timber panels, 86
Striking
  centres, 161
  considerations, 30
  damage, 27
  features, 26
  fillets, 131
  force, 27
  provision of draw, 27
  safety, 27, 29
  time, 28
  weight of form, by, 26, 131

Stripping fillets, 127
Subcontractors
　formwork, 15
　mould manufacture, 16
Supports
　cantilevers, to, 157
　easing, 157
　removal, 29
　standing props, 27, 162
Surface finish
　blemishes, 18
　effect of striking time, 29
　exposed aggregate, 187
　grades, 17
　grit blasting, 19

Tableform, 198
Taping joints, 18
Ties
　anchor loads, 250
　avoiding blocks, 68
　bolts, 31
　bright steel, 123
　buried ties, 79, 110, 129
　choice, 52
　economy, 253
　interrelation with supports, 52
　positioning, 123
　precasting, in, 177
　small shafts, in, 245
　system requirements, 244
　thickness of rods, 124
　tie rods, 78
　types available, 247
　typical arrangements, 246, 251
　waling, in, 123, 134
　waterproof, 252
Timber
　absorbency, 86
　advantages, 83
　feature fixing, 86
　manufacturing arrangements, 99

Timber—*contd.*
　number of uses, 85
　typical sizes, 84
Tolerances
　accuracy of precasting, 15
　mould sizes, 56
　relating to finished component, 26
Top forms
　combating uplift, 24
　I section beams, to, 23
　mix design, 24
Travellers, 47
Tunnel forms, 202

Vibration
　attachment, 23
　cave in, 17
　damage to sheathing, 88
　intensity in columns, 142
　nailing wedges, 108

Waffle formers
　floor casting, 161, 224
　grillages for, 163
　identification by colour, 163
Waling
　adjusting formwork, 121
　casting method, 122
　circular, 135
　conical, 135
　details, 120
　general construction, 133
　single-sided, 128, 129
　tapering, 137
　using prefabricated panels, 133
　wall forms, large, 195
Wedges, 117
Workmanship, 25, 26
Works erection, 232